I0061964

John Tyndall

The International Scientific Series

The Forms of Water in Clouds and Rivers, Ice and Glaciers

John Tyndall

The International Scientific Series
The Forms of Water in Clouds and Rivers, Ice and Glaciers

ISBN/EAN: 9783741143885

Manufactured in Europe, USA, Canada, Australia, Japa

Cover: Foto ©berggeist007 / pixelio.de

Manufactured and distributed by brebook publishing software
(www.brebook.com)

John Tyndall

The International Scientific Series

THE FORMS OF WATER

IN

CLOUDS & RIVERS, ICE & GLACIERS

BY

JOHN TYNDALL, LL.D. F.R.S.

PROFESSOR OF NATURAL PHILOSOPHY IN THE ROYAL INSTITUTION

*WITH THIRTY-FIVE ILLUSTRATIONS DRAWN AND ENGRAVED
UNDER THE DIRECTION OF THE AUTHOR*

THIRD EDITION

HENRY S. KING & Co.
65 CORNHILL & 12 PATERNOSTER ROW, LONDON
1873

.

AFTER an absence of twelve years, I visited the Mer de Glace last June. It exhibited in a striking degree that excess of consumption over supply which, if continued, will eventually reduce the Swiss glaciers to the mere spectres of their former selves. When I first saw the Mer de Glace its ice-cliffs towered over Les Mottets, and an arm of the Arveiron, issuing from the cliffs, plunged as a powerful cascade down the rocks. The ice has now shrunk far behind them. A huge moraine, left behind by the retreating glacier, will mark, for some time to come, its recent magnitude. The vault of the Arveiron has dwindled considerably. The way up to the Chapeau lies on the top of a lateral moraine, reached a few years ago by the surface of the glacier, the present surface lying far below. The visible and continual breaking away of the moraines, left thus stranded on the mountain flank, explains the absence of ancient ridges on the mountains where the slopes are steep. The ice-cascade of the Géant has suffered much from the general waste. Its crevasses are still wild, but the ice-cliffs and séracs of former days are but poorly represented to-day. The great Aletsch and its neighbours exhibit similar evidences of diminution. I found moreover this year that the two ancient moraines mentioned in paragraph 364 are parts of the same great lateral moraine which flanked the glacier for a long period, during which its magnitude must have remained practically constant. The place occupied by the ancient ice-river is rendered strikingly conspicuous by this well-preserved boundary.

During my residence at the Bel Alp this year, a catastrophe occurred which renders, for the time being, the description of the Märgelin See given in § 50 inappropriate. In company with two young friends I had descended the glacier and passed through the gorge of the Massa. On our return to the Bel Alp we found the domestics of the hotel leaning out of the windows and looking excitedly towards the glacier. From it proceeded a sound which resembled the roar of a cataract. The servants remarked that the Märgelin See must have broken loose. This was the case. For a time, however, the water flowed beneath the glacier; but at a point about midway between the Bel Alp and the

Æggischhorn, it broke forth on the Æggischhorn side, und formed a torrent between the glacier and the slope of the mountain. In some places this river was more than sixty yards wide, at others it was contracted to less than one-fifth of this width. Broken cascades of great height were formed here and there by successive ledges of ice, the torrent leaping with indescribable fury from ledge to ledge, and sending a smoke of spray into the air. At one place the bottom of the torrent was deep soft sand, which, after the water had passed, could be seen to have been tortured into huge funnels by the whirling eddies overhead.

Soon after we reached the Bel Alp, on the occasion just referred to, the front of the torrent appeared at the opposite side of the valley carrying everything movable before it, and immediately afterwards swept through the hollow that we had traversed a little earlier in the day. When at the end of the glacier I was struck by the force and volume of the Massa, and the grandeur of its vault, but I could not then account for the huge blocks of ice which it incessantly carried down. Doubtless the eruption above had been partial before the grand rush set in. The Rhone was considerably swollen, crops were damaged or ruined, and the driver of the diligence was sorely perplexed to find

himself in three feet of water, without any apparent
reason, on the public highway.   Two or three days
subsequently I learned at the Æggischhorn that an
engineer had been sent up to report on the possibility
of opening a channel, so as to prevent any future accumulation of water in the Mürgelin See.   If this be
done a useful end will be gained, by the abolition, however, of one of the most beautiful objects in Switzerland.

J. TYNDALL.

September 1872.

CONTENTS.

xii CONTENTS.

THE FORMS OF WATER

IN

CLOUDS AND RIVERS, ICE AND GLACIERS.

§ 1. *Clouds, Rains, and Rivers.*

1. EVERY occurrence in Nature is preceded by other
occurrences which are its causes, and succeeded by
others which are its effects. The human mind is not
satisfied with observing and studying any natural occurrence alone, but takes pleasure in connecting every
natural fact with what has gone before it, and with
what is to come after it.

2. Thus, when we enter upon the study of rivers and
glaciers, our interest will be greatly augmented by
taking into account not only their actual appearances,
but also their causes and effects.

3. Let us trace a river to its source. Beginning
where it empties itself into the sea, and following it

B

backwards, we find it from time to time joined by
tributaries which swell its waters. The river of course
becomes smaller as these tributaries are passed. It
shrinks first to a brook, then to a stream; this again
divides itself into a number of smaller streamlets,
ending in mere threads of water. These constitute
the source of the river, and are usually found among
hills.

4. Thus the Severn has its source in the Welsh
Mountains; the Thames in the Cotswold Hills; the
Danube in the hills of the Black Forest; the Rhine and
the Rhone in the Alps; the Ganges in the Himalaya
Mountains; the Euphrates near Mount Ararat; the
Garonne in the Pyrenees; the Elbe in the Giant Moun-
tains of Bohemia; the Missouri in the Rocky Mountains,
and the Amazon in the Andes of Peru.

5. But it is quite plain that we have not yet reached
the real beginning of the rivers. Whence do the
earliest streams derive their water? A brief residence
among the mountains would prove to you that they are
fed by rains. In dry weather you would find the
streams feeble, sometimes indeed quite dried up. In
wet weather you would see them foaming torrents. In
general these streams lose themselves as little threads
of water upon the hill sides; but sometimes you may
trace a river to a definite spring. The river Albula in
Switzerland, for instance, rushes at its origin in considerable volume from a mountain side. But you very

soon assure yourself that such springs are also fed by rain, which has percolated through the rocks or soil, and which, through some orifice that it has found or formed, comes to the light of day.

6. But we cannot end here. Whence comes the rain which forms the mountain streams? Observation enables you to answer the question. Rain does not come from a clear sky. It comes from clouds. But what are clouds? Is there nothing you are acquainted with which they resemble? You discover at once a likeness between them and the condensed steam of a locomotive. At every puff of the engine a cloud is projected into the air. Watch the cloud sharply: you notice that it first forms at a little distance from the top of the funnel. Give close attention and you will sometimes see a perfectly clear space between the funnel and the cloud. Through that clear space the thing which makes the cloud must pass. What, then, is this thing which at one moment is transparent and invisible, and at the next moment visible as a dense opaque cloud?

7. It is the *steam* or *vapour of water* from the boiler. Within the boiler this steam is transparent and invisible; but to keep it in this invisible state a heat would be required as great as that within the boiler. When the vapour mingles with the cold air above the hot funnel it ceases to be vapour. Every bit of steam shrinks, when chilled, to a much more minute particle of

water. The liquid particles thus produced form a kind of *water-dust* of exceeding fineness, which floats in the air, and is called *a cloud*.

8. Watch the cloud-banner from the funnel of a running locomotive; you see it growing gradually less dense. It finally melts away altogether, and if you continue your observations you will not fail to notice that the speed of its disappearance depends upon the character of the day. In humid weather the cloud hangs long and lazily in the air; in dry weather it is rapidly licked up. What has become of it? It has been reconverted into true invisible vapour.

9. The *drier* the air, and the *hotter* the air, the greater is the amount of cloud which can be thus dissolved in it. When the cloud first forms, its quantity is far greater than the air is able to maintain in an invisible state. But as the cloud mixes gradually with a larger mass of air it is more and more dissolved, and finally passes altogether from the condition of a finely-divided liquid into that of transparent vapour or gas.

10. Make the lid of a kettle air-tight, and permit the steam to issue from the pipe; a cloud is precipitated in all respects similar to that issuing from the funnel of the locomotive.

11. Permit the steam as it issues from the pipe to pass through the flame of a spirit-lamp, the cloud is instantly dissolved by the heat, and is not again precipitated. With a special boiler and a special nozzle the

experiment may be made more striking, but not more instructive, than with the kettle.

12. Look to your bedroom windows when the weather is very cold outside; they sometimes stream with water derived from the condensation of the aqueous vapour from your own lungs. The windows of railway carriages in winter show this condensation in a striking manner. Pour cold water into a dry drinking-glass on a summer's day: the outside surface of the glass becomes instantly dimmed by the precipitation of moisture. On a warm day you notice no vapour in front of your mouth, but on a cold day you form there a little cloud derived from the condensation of the aqueous vapour from the lungs.

13. You may notice in a ball-room that as long as the door and windows are kept closed, and the room remains hot, the air remains clear; but when the doors or windows are opened a dimness is visible, caused by the precipitation to fog of the aqueous vapour of the ball-room. If the weather be intensely cold the entrance of fresh air may even cause *snow* to fall. This has been observed in Russian ball-rooms; and also in the subterranean stables at Erzeroom, when the doors are opened and the cold morning air is permitted to enter.

14. Even on the driest day this vapour is never absent from our atmosphere. The vapour diffused through the air of this room may be congealed to hour frost in your presence. This is done by filling a

vessel with a mixture of pounded ice and salt, which
is colder than the ice itself, and which, therefore, con-
denses and freezes the aqueous vapour. The surface
of the vessel is finally coated with a frozen fur, so
thick that it may be scraped away and formed into a
snow-ball.

15. To produce the cloud, in the case of the locomotive and the kettle, *heat* is necessary. By heating
the water we first convert it into steam, and then by
chilling the steam we convert it into cloud. Is there
any fire in nature which produces the clouds of our
atmosphere? There is: the fire of the sun.

16. Thus, by tracing backward, without any break in
the chain of occurrences, our river from its end to its
real beginnings, we come at length to the sun.

§ 2.

17. There are, however, rivers which have sources
somewhat different from those just mentioned. They
do not begin by driblets on a hill side, nor can they be
traced to a spring. Go, for example, to the mouth of
the river Rhone, and trace it backwards to Lyons, where
it turns to the east. Bending round by Chambery, you
come at length to the Lake of Geneva, from which the
river rushes, and which you might be disposed to regard
as the source of the Rhone. But go to the head of the
lake, and you find that the Rhone there enters it, that
the lake is in fact a kind of expansion of the river.

Follow this upwards; you find it joined by smaller rivers from the mountains right and left. Pass these, and push your journey higher still. You come at length to a huge mass of ice—the end of a glacier—which fills the Rhone valley, and from the bottom of the glacier the river rushes. In the glacier of the Rhone you thus find the source of the river Rhone.

18. But again we have not reached the real beginning of the river. You soon convince yourself that this earliest water of the Rhone is produced by the melting of the ice. You get upon the glacier and walk upwards along it. After a time the ice disappears and you come upon snow. If you are a competent mountaineer you may go to the very top of this great snow-field, and if you cross the top and descend at the other side you finally quit the snow, and get upon another glacier called the Trift, from the end of which rushes a river smaller than the Rhone.

19. You soon learn that the mountain snow feeds the glacier. By some means or other the snow is converted into ice. But whence comes the snow? Like the rain, it comes from the clouds, which, as before, can be traced to vapour raised by the sun. Without solar fire we could have no atmospheric vapour, without vapour no clouds, without clouds no snow, and without snow no glaciers. Curious then as the conclusion may be, the cold ice of the Alps has its origin in the heat of the sun.

§ 3. *The Waves of Light.*

20. But what is the sun? We know its size and its
weight. We also know that it is a globe of fire far
hotter than any fire upon earth. But we now enter
upon another enquiry. We have to learn definitely
what is the meaning of solar light and solar heat; in
what way they make themselves known to our senses;
by what means they get from the sun to the earth, and
how, when there, they produce the clouds of our atmosphere, and thus originate our rivers and our glaciers.

21. If in a dark room you close your eyes and press
the eyelid with your finger-nail, a circle of light will be
seen opposite to the point pressed, while a sharp blow
upon the eye produces the impression of a flash of light.
There is a nerve specially devoted to the purposes of
vision which comes from the brain to the back of the
eye, and there divides into fine filaments, which are
woven together to a kind of screen called the *retina*.
The retina can be excited in various ways so as to produce the consciousness of light; it may, as we have
seen, be excited by the rude mechanical action of a blow
imparted to the eye.

22. There is no spontaneous creation of light by
the healthy eye. To excite vision the retina must be
affected by something coming from without. What is
that something? In some way or other luminous

bodies have the power of affecting the retina—but *how?*

23. It was long supposed that from such bodies issued, with inconceivable rapidity, an inconceivably fine matter, which flew through space, passed through the pores supposed to exist in the humours of the eye, reached the retina behind, and by their shock against the retina, aroused the sensation of light.

24. This theory, which was supported by the greatest men, among others by Sir Isaac Newton, was found competent to explain a great number of the phenomena of light, but it was not found competent to explain *all* the phenomena. As the skill and knowledge of experimenters increased, large classes of facts were revealed which could only be explained by assuming that light was produced, not by a fine matter flying through space and hitting the retina, but by the shock of minute *waves* against the retina.

25. Dip your finger into a basin of water, and cause it to quiver rapidly to and fro. From the point of disturbance issue small ripples which are carried forward by the water, and which finally strike the basin. Here, in the vibrating finger, you have a source of agitation; in the water you have a vehicle through which the finger's motion is transmitted, and you have finally the side of the basin which receives the shock of the little waves.

26. In like manner, according to the *wave theory* of

light, you have a source of agitation in the vibrating
atoms, or smallest particles, of the luminous body; you
have a vehicle of transmission in a substance which is
supposed to fill all space, and to be diffused through the
humours of the eye; and finally, you have the retina,
which receives the successive shocks of the waves.
These shocks are supposed to produce the sensation of
light.

27. We are here dealing, for the most part, with
suppositions and assumptions merely. We have never
seen the atoms of a luminous body, nor their motions.
We have never seen the medium which transmits their
motions, nor the waves of that medium. How, then, do
we come to assume their existence?

28. Before such an idea could have taken any real
root in the human mind, it must have been well disciplined and prepared by observations and calculations of
ordinary wave-motion. It was necessary to know how
both water-waves and sound-waves are formed and
propagated. It was above all things necessary to know
how waves, passing through the same medium, act upon
each other. Thus disciplined, the mind was prepared
to detect any resemblance presenting itself between the
action of light and that of waves. Great classes of
optical phenomena accordingly appeared which could
be accounted for in the most complete and satisfactory
manner by assuming them to be produced by waves, and
which could not be otherwise accounted for. It is

because of its competence to explain all the phenomena of light that the wave theory now receives universal acceptance on the part of scientific men. Let me use an illustration. We infer from the flint implements recently found in such profusion all over England and in other countries, that they were produced by men, and also that the Pyramids of Egypt were built by men, because, as far as our experience goes, nothing but men could form such implements or build such Pyramids. In like manner, we infer from the phenomena of light the agency of waves, because, as far as our experience goes, no other agency could produce the phenomena.

§ 4. *The Waves of Heat which produce the Vapour of our Atmosphere and melt our Glaciers.*

29. Thus, in a general way, I have given you the conception and the grounds of the conception, which regards light as the product of wave-motion; but we must go farther than this, and follow the conception into some of its details. We have all seen the waves of water, and we know they are of different sizes—different in length and different in height. When, therefore, you are told that the atoms of the sun, and of almost all other luminous bodies, vibrate at different rates, and produce waves of different sizes, your experience of water-waves will enable you to form a tolerably clear notion of what is meant.

30. As observed above, we have never seen the light-waves, but we judge of their presence, their position, and their magnitude, by their effects. Thei lengths have been thus determined, and found to vary from about $\frac{1}{76000}$th to $\frac{1}{30000}$th of an inch.

31. But besides those which produce light, the sun sends forth incessantly a multitude of waves which produce no light. The largest waves which the sun sends forth are of this non-luminous character, though they possess the highest heating power.

32. A common sunbeam contains waves of all kinds, but it is possible to *sift* or *filter* the beam so as to intercept all its light, and to allow its obscure heat to pass unimpeded. For substances have been discovered which, while intensely opaque to the light-waves, are almost perfectly transparent to the others. On the other hand, it is possible, by the choice of proper substances, to intercept in a great degree the pure heat-waves, and to allow the pure light-waves free transmission. This last separation is, however, not so perfect as the first.

33. We shall learn presently how to detach the one class of waves from the other class, and to prove that waves competent to light a fire, fuse metal, or burn the hand like a hot solid, may exist in a perfectly dark place.

34. Supposing, then, that we withdraw, in the first instance the large heat-waves, and allow the light-

waves alone to pass. These may be concentrated by
suitable lenses and sent into water without sensibly
warming it. Let the light-waves now be withdrawn,
and the larger heat-waves concentrated in the same
manner; they may be caused to boil the water almost
instantaneously.

35. This is the point to which I wished to lead you,
and which without due preparation could not be under-
stood. You now perceive the important part played by
these large darkness-waves, if I may use the term, in
the work of evaporation. When they plunge into seas,
lakes, and rivers, they are intercepted close to the surface, and they heat the water at the surface, thus
causing it to evaporate; the light-waves at the same
time entering to great depths without sensibly heating
the water through which they pass. Not only, therefore, is it the sun's fire which produces evaporation,
but a particular constituent of that fire, the existence
of which you probably were not aware of.

36. Further, it is these selfsame lightless waves
which, falling upon the glaciers of the Alps, melt the
ice and produce all the rivers flowing from the glaciers;
for I shall prove to you presently that the light-waves,
even when concentrated to the uttermost, are unable
to melt the most delicate hoar-frost; much less would
they be able to produce the copious liquefaction observed
upon the glaciers.

37. These large lightless waves of the sun, as well as

the heat-waves issuing from non-luminous hot bodies,
are frequently called obscure or invisible heat.

We have here an example of the manner in which
phenomena, apparently remote, are connected together
in this wonderful system of things that we call Nature.
You cannot study a snow-flake profoundly without being
led back by it step by step to the constitution of the sun.
It is thus throughout Nature. All its parts are inter-
dependent, and the study of any one part *completely*
would really involve the study of all.

§ 5. *Experiments to prove the foregoing Statements.*

38. Heat issuing from any source not visibly red
cannot be concentrated so as to produce the intense
effects just referred to. To produce these it is necessary to employ the obscure heat of a body raised to the
highest possible state of incandescence. The sun is
such a body, and its dark heat is therefore suitable
for experiments of this nature.

39. But in the atmosphere of London, and for experiments such as ours, the heat-waves emitted by coke
raised to intense whiteness by a current of electricity
are much more manageable than the sun's waves.
The electric light has also the advantage that its dark
radiation embraces a larger proportion of the total radiation than the dark heat of the sun. In fact, the force
or energy, if I may use the term, of the dark waves of
the electric light is fully seven times that of its light-

waves. The electric light, therefore, shall be employed in our experimental demonstrations.

40. From this source a powerful beam is sent through the room, revealing its track by the motes floating in the air of the room; for were the motes entirely absent the beam would be unseen. It falls upon a concave mirror (a glass one silvered behind will answer) and is gathered up by the mirror into a cone of reflected rays; the luminous apex of the cone, which is the *focus* of the mirror, being about fifteen inches distant from its reflecting surface. Let us mark the focus accurately by a pointer.

41. And now let us place in the path of the beam a substance perfectly opaque to light. This substance is iodine dissolved in a liquid called bisulphide of carbon. The light at the focus instantly vanishes when the dark solution is introduced. But the solution is intensely transparent to the dark waves, and a focus of such waves remains in the air of the room after the light has been abolished. You may feel the heat of these waves with your hand; you may let them fall upon a thermometer, and thus prove their presence; or, best of all, you may cause them to produce a current of electricity, which deflects a large magnetic needle. The magnitude of the deflection is a measure of the heat.

42. Our object now is, by the use of a more powerful lamp, and a better mirror (one silvered in front and with a shorter focal distance), to intensify the

uction here rendered so sensible. As before, the focus
is rendered strikingly visible by the intense illumina-
tion of the dust particles. We will first filter the
beam so as to intercept its dark waves, and then permit the purely luminous waves to exert their utmost
power on a small bundle of gun-cotton placed at the
focus.

43. No effect whatever is produced. The gun-cotton
might remain there for a week without ignition. Let
us now permit the unfiltered beam to act upon the
cotton. It is instantly dissipated in an explosive flash.
This experiment proves that the light-waves are incompetent to explode the cotton, while the waves of the
full beam are competent to do so; hence we may conclude that the dark waves are the real agents in the
explosion.

44. But this conclusion would be only probable; for
it might be urged that the *mixture* of the dark waves
and the light-waves is necessary to produce the result.
Let us then, by means of our opaque solution, isolate
our dark waves and converge them on the cotton. It
explodes as before.

45. Hence it is the dark waves, and they only, that
are concerned in the ignition of the cotton.

46. At the same dark focus sheets of platinum are
raised to vivid redness; zinc is burnt up; paper instantly blazes; magnesium wire is ignited; charcoal
within a receiver containing oxygen is set burning; a

diamond similarly placed is caused to glow like a star, being afterwards gradually dissipated. And all this while the *air* at the focus remains as cool as in any other part of the room.

47. To obtain the light-waves we employ a clear solution of alum in water; to obtain the dark waves we employ the solution of iodine above referred to. But as before stated (32), the alum is not so perfect a filter as the iodine; for it transmits a portion of the obscure heat.

48. Though the light-waves here prove their incompetence to ignite gun-cotton, they are able to burn up black paper; or, indeed, to explode the cotton when it is blackened. The white cotton does not absorb the light, and without absorption we have no heating. The blackened cotton absorbs, is heated, and explodes.

49. Instead of a solution of alum, we will employ for our next experiment a cell of pure water, through which the light passes without sensible absorption. At the focus is placed a test-tube also containing water, the full force of the light being concentrated upon it. The water is not sensibly warmed by the concentrated waves. We now remove the cell of water; no change is visible in the beam, but the water contained in the test-tube now boils.

50. The light-waves being thus proved ineffectual, and the full beam effectual, we may infer that it is the dark waves that do the work of heating. But we clench

c

our inference by employing our opaque iodine filter.
Placing it on the path of the beam, the light is entirely
stopped, but the water boils exactly as it did when the
full beam fell upon it.

51. The truth of the statement made in paragraph
(34) is thus demonstrated.

52. And now with regard to the melting of ice. On
the surface of a flask containing a freezing mixture
we obtain a thick fur of hoar-frost. Sending the
beam through a water-cell, its luminous waves are concentrated upon the surface of the flask. Not a spicula
of the frost is dissolved. We now remove the water-cell,
and in a moment a patch of the frozen fur as large as
half-a-crown is melted. Hence, inasmuch as the full
beam produces this effect, and the luminous part of the
beam does not produce it, we fix upon the dark portion
the melting of the frost.

53. As before, we clench this inference by concentra-
ting the dark waves alone upon the flask. The frost is
dissipated exactly as it was by the full beam.

54. These effects are rendered strikingly visible by
darkening with ink the freezing mixture within the
flask. When the hoar frost is removed, the blackness
of the surface from which it had been melted comes
out in strong contrast with the adjacent snowy whiteness. When the flask itself, instead of the freezing
mixture, is blackened, the purely luminous waves being
absorbed by the glass, warm it; the glass reacts upon

the frost, and melts it. Hence the wisdom of darkening, instead of the flask itself, the mixture within the flask.

55. This experiment proves to demonstration the statement in paragraph (36): that it is the dark waves of the sun that melt the mountain snow and ice, and originate all the rivers derived from glaciers.

There are writers who seem to regard science as an aggregate of facts, and hence doubt its efficacy as an exercise of the reasoning powers. But all that I have here taught you is the result of reason, taking its stand, however, upon the sure basis of observation and experiment. And this is the spirit in which our further studies are to be pursued.

§ 6. *Oceanic Distillation.*

56. The sun, you know, is never exactly overhead in England. But at the equator, and within certain limits north and south of it, the sun at certain periods of the year is directly overhead at noon. These limits are called the Tropics of Cancer and of Capricorn. Upon the belt comprised between these two circles the sun's rays fall with their mightiest power; for here they shoot directly downwards, and heat both earth and sea more than when they strike slantingly.

57. When the vertical sunbeams strike the land they heat it, and the air in contact with the hot soil becomes heated in turn. But when heated the air expands, and when it expands it becomes lighter. This lighter air

c 2

rises, like wood plunged into water, through the heavier air overhead.

58. When the sunbeams fall upon the sea the water is warmed, though not so much as the land. The warmed water expands, becomes thereby lighter, and therefore continues to float upon the top. This upper layer of water warms to some extent the air in contact with it, but it also sends up a quantity of aqueous vapour, which being far lighter than air, helps the latter to rise. Thus both from the land and from the sea we have ascending currents established by the action of the sun.

59. When they reach a certain elevation in the atmosphere, these currents divide and flow, part towards the north and part towards the south; while from the north and the south a flow of heavier and colder air sets in to supply the place of the ascending warm air.

60. Incessant circulation is thus established in the atmosphere. The equatorial air and vapour flow above towards the north and south poles, while the polar air flows below towards the equator. The two currents of air thus established are called the upper and the lower trade winds.

61. But before the air returns from the poles great changes have occurred. For the air as it quitted the equatorial regions was laden with aqueous vapour, which could not subsist in the cold polar regions. It is

there precipitated, falling sometimes as rain, or more commonly as snow. The land near the pole is covered with this snow, which gives birth to vast glaciers in a manner hereafter to be explained.

62. It is necessary that you should have a perfectly clear view of this process, for great mistakes have been made regarding the manner in which glaciers are related to the heat of the sun.

63. It was supposed that if the sun's heat were diminished, greater glaciers than those now existing would be produced. But the lessening of the sun's heat would infallibly diminish the quantity of aqueous vapour, and thus cut off the glaciers at their source. A brief illustration will complete your knowledge here.

64. In the process of ordinary distillation, the liquid to be distilled is heated and converted into vapour in one vessel, and chilled and reconverted into liquid in another. What has just been stated renders it plain that the earth and its atmosphere constitute a vast distilling apparatus in which the equatorial ocean plays the part of the boiler, and the chill regions of the poles the part of the condenser. In this process of distillation *heat* plays quite as necessary a part as *cold*, and before Bishop Heber could speak of ' Greenland's icy mountains,' the equatorial ocean had to be warmed by the sun. We shall have more to say upon this question afterwards.

2
THE FORMS OF WATER IN

ILLUSTRATIVE EXPERIMENTS.

65. I have said that when heated, air expands. If
you wish to verify this for yourself, proceed thus. Take
an empty flask, stop it by a cork; pass through the cork
a narrow glass tube. By heating the tube in a spirit-
lamp you can bend it downwards, so that when the
flask is standing upright the open end of the narrow
tube may dip into water. Now cause the flame of
your spirit-lamp to play against the flask. The flame
heats the glass, the glass heats the air; the air expands, is driven through the narrow tube, and issues
in a storm of bubbles from the water.

66. Were the heated air unconfined, it would rise in
the heavier cold air. Allow a sunbeam or any other
intense light to fall upon a white wall or screen in a dark
room. Bring a heated poker, a candle, or a gas flame
underneath the beam. An ascending current rises from
the heated body through the beam, and the action
of the air upon the light is such as to render the
wreathing and waving of the current strikingly visible
upon the screen. When the air is hot enough, and therefore light enough, if entrapped in a paper bag it carries
the bag upwards, and you have the Fire-balloon.

67. Fold two sheets of paper into two cones and
suspend them with their closed points upwards from
the end of a delicate balance. See that the cones

balance each other. Then place for a moment the flame of a spirit-lamp beneath the open base of one of them ; the hot air ascends from the lamp and instantly tosses upwards the cone above it.

68. Into an inverted glass shade introduce a little smoke. Let the air come to rest, and then simply place your hand at the open mouth of the shade. Mimic hurricanes are produced by the air warmed by the hand, which are strikingly visible when the smoke is illuminated by a strong light.

69. The heating of the tropical air by the sun is *indirect.* The solar beams have scarcely any power to heat the air through which they pass ; but they heat the land and ocean, and these communicate their heat to the air in contact with them. The air and vapour start upwards charged with the heat thus communicated.

§ 7. *Tropical Rains.*

70. But long before the air and vapour from the equator reach the poles, precipitation occurs. Wherever a humid warm wind mixes with a cold dry one, rain falls. Indeed the heaviest rains occur at those places where the sun is vertically overhead. We must enquire a little more closely into their origin.

71. Fill a bladder about two-thirds full of air at the sea level, and take it to the summit of Mont Blanc. As you ascend, the bladder becomes more and more dis·

tended; at the top of the mountain it is fully distended, and has evidently to bear a pressure from within. Returning to the sea level you find that the tightness disappears, the bladder finally appearing as flaccid as at first.

72. The reason is plain. At the sea level the air within the bladder has to bear the pressure of the whole atmosphere, being thereby squeezed into a comparatively small volume. In ascending the mountain, you leave more and more of the atmosphere behind; the pressure becomes less and less, and by its expansive force the air within the bladder swells as the outside pressure is diminished. At the top of the mountain the expansion is quite sufficient to render the bladder tight, the pressure within being then actually greater than the pressure without. By means of an air-pump we can show the expansion of a balloon partly filled with air, when the external pressure has been in part removed.

73. But why do I dwell upon this? Simply to make plain to you that the *unconfined air*, heated at the earth's surface, and ascending by its lightness, must expand more and more the higher it rises in the atmosphere.

74. And now I have to introduce to you a new fact, towards the statement of which I have been working for some time. It is this:—*The ascending air is chilled by its expansion.* Indeed this chilling is one source of the coldness of the higher atmospheric regions. And

now fix your eye upon those mixed currents of air and aqueous vapour which rise from the warm tropical ocean. They start with plenty of heat to preserve the vapour as vapour; but as they rise they come into regions already chilled, and they are still further chilled by their own expansion. The consequence might be foreseen. The load of vapour is in great part precipitated, dense clouds are formed, their particles coalesce to rain-drops, which descend daily in gushes so profuse that the word 'torrential' is used to express the copiousness of the rain-fall. I could show you this chilling by expansion, and also the consequent precipitation of clouds.

75. Thus long before the air from the equator reaches the poles its vapour is in great part removed from it, having redescended to the earth as rain. Still a good quantity of the vapour is carried forward, which yields hail, rain, and snow in northern and southern lands.

ILLUSTRATIVE EXPERIMENTS.

76. I have said that the air is chilled during its expansion. Prove this, if you like, thus. With a condensing syringe, you can force air into an iron box furnished with a stopcock, to which the syringe is screwed. Do so till the density of the air within the box is doubled or trebled. Immediately after this condensation, both the box and the air within it are *warm*, and can be proved to be so by a proper thermometer.

Simply turn the cock and allow the compressed air to stream into the atmosphere. The current, if allowed to strike a thermometer, visibly chills it; and with other instruments the chill may be made more evident still. Even the hand feels the chill of the expanding air.

77. Throw a strong light, a concentrated sunbeam for example, across the issuing current; if the compressed air be ordinary humid air, you see the precipitation of a little cloud by the chill accompanying the expansion. This cloud-formation may, however, be better illustrated in the following way:—

78. In a darkened room send a strong beam of light through a glass tube three feet long and three inches wide, stopped at its ends by glass plates. Connect the tube by means of a stopcock with a vessel of about one-fourth its capacity, from which the air has been removed by an air-pump. The exhausted cylinder of the pump itself will answer capitally. Fill the glass tube with humid air; then simply turn on the stopcock which connects it with the exhausted vessel. Having more room the air expands, cold accompanies the expansion, and, as a consequence, a dense and brilliant cloud immediately fills the tube. If the experiment be made for yourself alone you may see the cloud in ordinary daylight; indeed, the brisk exhaustion of any receiver filled with humid air is known to produce this condensation.

79. Other vapours than that of water may be thus

precipitated, some of them yielding clouds of intense brilliancy, and displaying iridescences, such as are sometimes, but not frequently, seen in the clouds floating over the Alps.

80. In science what is true for the small is true for the large. Thus by combining the conditions observed on a large scale in nature we obtain on a small scale the phenomena of atmospheric clouds.

§ 8. *Mountain Condensers.*

81. To complete our view of the process of atmospheric precipitation we must take into account the action of mountains. Imagine a south-west wind blowing across the Atlantic towards Ireland. In its passage it charges itself with aqueous vapour. In the south of Ireland it encounters the mountains of Kerry: the highest of these is Magillicuddy's Reeks, near Killarney. Now the lowest stratum of this Atlantic wind is that which is most fully charged with vapour. When it encounters the base of the Kerry mountains it is tilted up and flows bodily over them. Its load of vapour is therefore carried to a height, it expands on reaching the height, it is chilled in consequence of the expansion, and comes down in copious showers of rain. From this, in fact, arises the luxuriant vegetation of Killarney; to this, indeed, the lakes owe their water supply. The cold crests of the mountains also aid in the work of condensation.

82. Note the consequence. There is a town called Cahirciveen to the south-west of Magillicuddy's Rocks, at which observations of the rainfall have been made, and a good distance further to the north-east, right in the course of the south-west wind, there is another town, called Portarlington, at which observations of rainfall have also been made. But before the wind reaches the latter station it has passed over the mountains of Kerry and left a great portion of its moisture behind it. What is the result? At Cahirciveen, as shown by Dr. Lloyd, the rainfall amounts to 59 inches in a year, while at Portarlington it is only 21 inches.

83. Again, you may sometimes descend from the Alps when the fall of rain and snow is heavy and incessant, into Italy, and find the sky over the plains of Lombardy blue and cloudless, the wind at the same time *blowing over the plain towards the Alps*. Below the wind is hot enough to keep its vapour in a perfectly transparent state; but it meets the mountains, is tilted up, expanded, and chilled. The cold of the higher summits also helps the chill. The consequence is that the vapour is precipitated as rain or snow, thus producing bad weather upon the heights, while the plains below, flooded with the same air, enjoy the aspect of the unclouded summer sun. Clouds blowing *from* the Alps are also sometimes dissolved over the plains of Lombardy.

84. In connection with the formation of clouds by mountains, one particularly instructive effect may be

here noticed. You frequently see a streamer of cloud
many hundred yards in length drawn out from an
Alpine peak. Its steadiness appears perfect, though a
strong wind may be blowing at the same time over the
mountain head. Why is the cloud not blown away?
It *is* blown away; its permanence is only apparent. At
one end it is incessantly dissolved, at the other end it
is incessantly renewed: supply and consumption being
thus equalized, the cloud appears as changeless as the
mountain to which it seems to cling. When the red
sun of the evening shines upon these cloud-streamers
they resemble vast torches with their flames blown
through the air.

§ 9. *Architecture of Snow.*

85. We now resemble persons who have climbed a
difficult peak, and thereby earned the enjoyment of a
wide prospect. Having made ourselves masters of the
conditions necessary to the production of mountain
snow, we are able to take a comprehensive and intelli-
gent view of the phenomena of glaciers.

86. A few words are still necessary as to the formation of snow. The molecules and atoms of all substances, when allowed free play, build themselves into
definite and, for the most part, beautiful forms called
crystals. Iron, copper, gold, silver, lead, sulphur, when
melted and permitted to cool gradually, all show this
crystallizing power. The metal bismuth shows it in a

particularly striking manner, and when properly fused and solidified, self-built crystals of great size and beauty are formed of this metal.

87. If you dissolve saltpetre in water, and allow the solution to evaporate slowly, you may obtain large crystals, for no portion of the salt is converted into vapour. The water of our atmosphere is fresh though it is derived from the salt sea. Sugar dissolved in water, and permitted to evaporate, yields crystals of sugar-candy. Alum readily crystallizes in the same way. Flints dissolved, as they sometimes are in nature, and permitted to crystallize, yield the prisms and pyramids of rock crystal. Chalk dissolved and crystallized yields Iceland spar. The diamond is crystallized carbon. All our precious stones, the ruby, sapphire, beryl, topaz, emerald, are all examples of this crystallizing power.

88. You have heard of the force of gravitation, and you know that it consists of an attraction of every particle of matter for every other particle. You know that planets and moons are held in their orbits by this attraction. But gravitation is a very simple affair compared to the force, or rather forces, of crystallization. For here the ultimate particles of matter, inconceivably small as they are, show themselves possessed of attractive and repellent poles, by the mutual action of which the shape and structure of the crystal are determined. In the solid condition the attracting poles are rigidly locked together; but if sufficient heat be applied the

bond of union is dissolved, and in the state of fusion the poles are pushed so far asunder as to be practically out of each other's range.   The natural tendency of the molecules to build themselves together is thus neutralized.

89. This is the case with water, which as a liquid is to all appearance formless.   When sufficiently cooled the molecules are brought within the play of the crystallizing force, and they then arrange themselves in forms of indescribable beauty.   When snow is produced in calm air, the icy particles build themselves into beautiful stellar shapes, each star possessing six rays. There is no deviation from this type, though in other respects the appearances of the snow-stars are infinitely various.   In the polar regions these exquisite forms were observed by Dr. Scoresby, who gave numerous drawings of them.   I have observed them in mid-winter filling the air, and loading the slopes of the Alps.   But in England they are also to be seen, and no words of mine could convey so vivid an impression of their beauty as the annexed drawings of a few of them, executed at Greenwich by Mr. Glaisher.

90. It is worth pausing to think what wonderful work is going on in the atmosphere during the formation and descent of every snow-shower: what building power is brought into play! and how imperfect seem the productions of human minds and hands when compared with those formed by the blind forces of nature!

91. But who ventures to call the forces of nature

blind?   In reality, when we speak thus we are describing
our own condition.   The blindness is ours; and what
we really ought to say, and to confess, is that our powers
are absolutely unable to comprehend either the origin
or the end of the operations of nature.

92. But while we thus acknowledge our limits, there
is also reason for wonder at the extent to which science
has mastered the system of nature. From age to age,
and from generation to generation, fact has been added
to fact, and law to law, the true method and order of
the Universe being thereby more and more revealed.
In doing this science has encountered and overthrown
various forms of superstition and deceit, of credulity
and imposture. But the world continually produces
weak persons and wicked persons; and as long as they
continue to exist side by side, as they do in this our
day, very debasing beliefs will also continue to infest
the world.

§ 10. *Atomic Poles.*

93. 'What did I mean when, a few moments ago
(88), I spoke of attracting and repellent poles?' Let
me try to answer this question. You know that astro-
nomers and geographers speak of the earth's poles, and
you have also heard of magnetic poles, the poles of a
magnet being the points at which the attraction and
repulsion of the magnet are as it were concentrated.

94. Every magnet possesses two such poles; and if

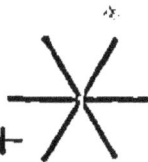

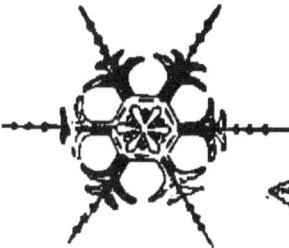

SNOW CRYSTALS.

D

iron filings be scattered over a magnet, each particle becomes also endowed with two poles. Suppose such particles devoid of weight and floating in our atmosphere, what must occur when they come near each other? Manifestly the repellent poles will retreat from each other, while the attractive poles will approach and finally lock themselves together. And supposing the particles, instead of a single pair, to possess several pairs of poles arranged at definite points over their surfaces; you can then picture them, in obedience to their mutual attractions and repulsions, building themselves together to form masses of definite shape and structure.

95. Imagine the molecules of water in calm cold air to be gifted with poles of this description, which compel the particles to lay themselves together in a definite order, and you have before your mind's eye the unseen architecture which finally produces the visible and beautiful crystals of the snow. Thus our first notions and conceptions of poles are obtained from the sight of our eyes in looking at the effects of magnetism; and we then transfer these notions and conceptions to particles which no eye has ever seen. The power by which we thus picture to ourselves effects beyond the range of the senses is what philosophers call the Imagination, and in the effort of the mind to seize upon the unseen architecture of crystals, we have an example of the 'scientific use' of this faculty. Without imagination we might have *critical* power, but not *creative* power in science.

§ 11. *Architecture of Lake Ice.*

96. We have thus made ourselves acquainted with the beautiful snow-flowers self-constructed by the molecules of water in calm cold air. Do the molecules show this architectural power when ordinary water is frozen? What, for example, is the structure of the ice over which we skate in winter? Quite as wonderful as the flowers of the snow. The observation is rare, if not new, but I have seen in water slowly freezing six-rayed ice-stars formed, and floating free on the surface. A six-rayed star, moreover, is typical of the construction of all our lake ice. It is built up of such forms wonderfully interlaced.

97. Take a slab of lake ice and place it in the path of a concentrated sunbeam. Watch the track of the beam through the ice. Part of the beam is stopped, part of it goes through; the former produces internal liquefaction, the latter has no effect whatever upon the ice. But the liquefaction is not uniformly diffused. From separate spots of the ice little shining points are seen to sparkle forth. Every one of those points is surrounded by a beautiful liquid flower with six petals.

98. Ice and water are so optically alike that unless the light fall properly upon these flowers you cannot see them. But what is the central spot? A vacuum. Ice swims on water because, bulk for bulk, it is lighter

than water; so that when ice is melted it shrinks in size. Can the liquid flowers then occupy the whole space of the ice melted? Plainly no. A little empty space is formed with the flowers, and this space, or rather its surface, shines in the sun with the lustre of burnished silver.

99. In all cases the flowers are formed parallel to the surface of freezing. They are formed when the sun shines upon the ice of every lake; sometimes in myriads, and so small as to require a magnifying glass to see them. They are always attainable, but their beauty is often marred by internal defects of the ice. Even one portion of the same piece of ice may show them exquisitely, while a second portion shows them imperfectly.

100. Annexed is a very imperfect sketch of these beautiful figures.

101. Here we have a reversal of the process of crystallization. The searching solar beam is delicate enough to take the molecules down without deranging the order of their architecture. Try the experiment for yourself with a pocket-lens on a sunny day. You will not find the flowers confused; they all lie parallel to the surface of freezing. In this exquisite way every bit of the ice over which our skaters glide in winter is put together.

102. I said in (97) that a portion of the sunbeam was stopped by the ice and liquefied it. What is this

portion? The dark heat of the sun. The great body of the light waves and even a portion of the dark ones,

LIQUID FLOWERS IN LAKE ICE.

pass through the ice without losing any of their heating power. When properly concentrated on combustible bodies, even after having passed through the ice, their burning power becomes manifest.

103. And the ice itself may be employed to concentrate them. With an ice-lens in the polar regions Dr. Scoresby has often concentrated the sun's rays so as to make them burn wood, fire gunpowder, and melt lead; thus proving that the heating power is retained by the rays, even after they have passed through so cold a substance.

104. By rendering the rays of the electric lamp parallel, and then sending them through a lens of ice, we obtain all the effects which Dr. Scoresby obtained with the rays of the sun.

§ 12. *The Source of the Arveiron. Ice Pinnacles, Towers,
and Chasms of the Glacier des Bois. Passage
to the Montanvert.*

105. Our preparatory studies are for the present
ended, and thus informed, let us approach the Alps.
Through the village of Chamouni, in Savoy, a river
rushes which is called the Arve. Let us trace this
river backwards from Chamouni. At a little distance
from the village the river forks ; one of its branches still
continues to be called the Arve, the other is the Arveiron. Following this latter we come to what is called
the ' source of the Arveiron '—a short hour's walk from
Chamouni. Here, as in the case of the Rhone already
referred to, you are fronted by a huge mass of ice, the
end of a glacier, and from an arch in the ice the Arveiron issues. Do not trust the arch in summer. Its
roof falls at intervals with a startling crash, and would
infallibly crush any person on whom it might fall.

106. We must now be observant. Looking about us
here, we find in front of the ice curious heaps and ridges
of débris, which are more or less concentric. These are
the *terminal moraines* of the glacier. We shall examine
them subsequently.

107. We now turn to the left, and ascend the slope
beside the glacier. As we ascend we get a better view,
and find that the ice here fills a narrow valley. We

come upon another singular ridge, not of fresh débris, like those lower down, but covered in part with trees, and appearing to be literally as 'old as the hills.' It tells a wonderful tale. We soon satisfy ourselves that the ridge is an ancient moraine, and at once conclude that the glacier, at some former period of its existence,

SOURCE OF THE ARVEIRON.

was vastly larger than it is now. This old moraine stretches right across the main valley, and abuts against the mountains at the opposite side.

108. Having passed the terminal portion of the glacier, which is covered with stones and rubbish, we find ourselves beside a very wonderful exhibition of ice.

The glacier descends a steep gorge, and in doing so is
riven and broken in the most extraordinary manner.
Here are towers, and pinnacles, and fantastic shapes
wrought out by the action of the weather, which put
one in mind of rude sculpture. Annexed is a sketch
of an ice-pinnacle. . From deep chasms in the glacier

ICE-PINNACLE.

issues a delicate shimmer of blue light. At times we
hear a sound like thunder, which arises either from
the falling of a tower of ice, or from the tumble of a
huge stone into a chasm. The glacier maintains this
wild and chaotic character for some time; and the best

iceman would find himself defeated in any attempt to get along it.

109. We reach a place called the Chapeau, where, if we wish, we can have refreshment in a little mountain hut. We then pass the *Mauvais Pas*, a precipitous rock, on the face of which steps are hewn, and the un-practised traveller is assisted by a rope. We pursue our journey, partly along the mountain side, and partly along a ridge of singularly artificial aspect—a *lateral moraine*. We at length face a house perched upon an eminence at the opposite side of the glacier. This is the auberge of the Montanvert, well known to all visitors to this portion of the Alps.

110. Here we cross the glacier. I should have told you that its lower part, including the broken portion we have passed, is called the Glacier des Bois; while the place that we are now about to cross is the beginning of the Mer de Glace. You feel that this term is not quite appropriate, for the glacier here is much more like a *river* of ice than a sea. The valley which it fills is about half a mile wide.

111. The ice may be riven where we enter upon it, but with the necessary care there is no difficulty in crossing this portion of the Mer de Glace. The clefts and chasms in the ice are called *crevasses*; we shall make their acquaintance on a grander scale by and by.

112. Look up and down this side of the glacier. It is considerably riven, but as we advance the crevasses

THE NEW CLIFFS SHOWED FROM OUR LEVEL AND THE GRANDE DE JORNADA, WITH OUR CAMP ABOUT THREEFOURTHS TO THE RIGHT.

will diminish, and we shall find very few of them at
the other side. Note this for future use. The ice is at
first dirty; but the dirt soon disappears, and you come
upon the clean crisp surface of the glacier. You have
already noticed that the clean ice is white, and that
from a distance it resembles snow rather than ice. This
is caused by the breaking up of the surface by the
solar heat. When you pound transparent rock-salt
into powder it is as white as table-salt, and it is the
minute fissuring of the surface of the glacier by the
sun's rays that causes it to appear white. *Within* the
glacier the ice is transparent. After an exhilarating
passage we get upon the opposite lateral moraine, and
ascend the steep slope from it to the Montanvert Inn.

§ 13. *The Mer de Glace and its Sources.   Our First
Climb to the Cleft Station.*

119. Here the view before us is very grand. We
look across the glacier at the beautiful pyramid of the
Aiguille du Dru (shown in our frontispiece); and to the
right at the Aiguille des Charmoz, with its sharp pinnacles bent as if they were ductile. Looking straight
up the glacier the view is bounded by the great crests
called La Grande Jorasse, nearly 14,000 feet high.
Our object now is to get into the very heart of the
mountains, and to pursue to its origin the wonderful
frozen river which we have just crossed.

114. Starting from the Montanvert with the glacier below us to our left, we soon reach some rocks resembling the Mauvais Pas; they are called *les Ponts*. We cross them and reach *l'Angle*, where we quit the land for the ice. We walk up the glacier, but before reaching the promontory called Trélaporte, we take once more to the mountain side; for though the path here has been forsaken on account of its danger, for the sake of knowledge we are prepared to incur danger to a reasonable extent. A little glacier reposes on the slope to our right. We may see a huge boulder or two poised on the end of the glacier, and, if fortunate, also see the boulder liberated and plunging violently down the slope. Presence of mind is all that is necessary to render our safety certain; but travellers do not always show presence of mind, and hence the path which formerly led over this slope has been forsaken. The whole slope is cumbered by masses of rock which this little glacier has sent down. These I wished you to see; by and by they shall be fully accounted for.

115. Above Trélaporte to the right you see a most singular cleft in the rocks, in the middle of which stands an isolated pillar, hewn out by the weather. Our next object is to get to the tower of rock to the left of that cleft, for from that position we shall gain a most commanding and instructive view of the Mer de Glace and its sources.

116. The cleft referred to, with its pillar, may be

seen to the right of the preceding engraving of the Mer
de Glace. Below the cleft is also seen the little glacier
just referred to.

117. We may reach this cleft by a steep gully, visible
from our present position, and leading directly up
to the cleft. But these gullies, or couloirs, are very
dangerous, being the pathways of stones falling from
the heights. We will therefore take the rocks to the
left of the gully, by close inspection ascertain their assailable points, and there attack them. In the Alps
as elsewhere wonderful things may be done by looking
steadfastly at difficulties, and testing them wherever
they appear assailable. We thus reach our station,
where the glory of the prospect, and the insight that
we gain as to the formation of the Mer de Glace, far
more than repay us for the labour of our ascent.

118. For we see the glacier below us, stretching its
frozen tongue downwards past the Montanvert. And
we now find this single glacier branching out into three
others, some of them wider than itself. Regard the
branch to the right, the Glacier du Géant. It stretches
smoothly up for a long distance, then becomes disturbed,
and then changes to a great frozen cascade, down which
the ice appears to tumble in wild confusion. Above
the cascade you see an expanse of shining snow, oc-
cupying an area of some square miles.

§ 14. *Ice-cascade and Snows of the Col du Géant.*

119. Instead of climbing to the height where we now stand, we might have continued our walk upon the Mer de Glace, turned round the promontory of Trélaporte, and walked right up the Glacier du Géant. We should have found ice under our feet up to the bottom of the cascade. It is not so compact as the ice lower down, but you would not think of refusing to call it ice.

120. As we approach the fall, the smooth and unbroken character of the glacier changes more and more. We encounter transverse ridges succeeding each other with augmenting steepness. The ice becomes more and more fissured and confused. We wind through tortuous ravines, climb huge ice-mounds, and creep cautiously along crumbling crests, with crevasses right and left. The confusion increases until further advance along the centre of the glacier is impossible.

121. But with the aid of an axe to cut steps in the steeper ice-walls and slopes we might, by swerving to either side of the glacier, work our way to the top of the cascade. If we ascended to the right, we should have to take care of the ice avalanches which sometimes thunder down the slopes; if to the left, we should have to take care of the stones let loose from the Aiguille Noire. After we had cleared the cascade, we should have to beware for a time of the crevasses, which for

some distance above the fall yawn terribly. But by
caution we could get round them, and sometimes cross
them by bridges of snow. Here the skill and knowledge to be acquired only by long practice come into
play; and here also the use of the Alpine rope suggests itself. For not only are the snow bridges often
frail, but whole crevasses are sometimes covered, the
unhappy traveller being first made aware of their existence by the snow breaking under his feet. Many lives
have thus been lost, and some quite recently.

122. Once upon the plateau above the ice-fall we
find the surface totally changed. Below the fall we
walked upon ice; here we are upon snow. After a
gentle but long ascent we reach a depression of the
ridge which bounds the snow-field at the top, and now
look over Italy. We stand upon the famous Col du
Géant.

123. They were no idle scamperers on the mountains
that made these wild recesses first known; it was not
the desire for health which now brings some, or the
desire for grandeur and beauty which brings others, or
the wish to be able to say that they have climbed a
mountain or crossed a col, which I fear brings a good
many more; it was a desire for *knowledge* that brought
the first explorers here, and on this col the celebrated
De Saussure lived for seventeen days, making scientific
observations.

§ 15. *Questioning the Glaciers.*

124. I would now ask you to consider for a moment
the facts which such an excursion places in our posses-
sion. The snow through which we have in idea trudged
is the snow of last winter and spring. Had we placed
last August a proper mark upon the surface of the
snow, we should find it this August at a certain depth
beneath the surface. A good deal has been melted
by the summer sun, but a good deal of it remains, and
it will continue until the snows of the coming winter
fall and cover it. This again will be in part preserved
till next August, a good deal of it remaining until it is
covered by the snow of the subsequent winter. We thus
arrive at the certain conclusion that on the plateau
of the Col du Géant *the quantity of snow that falls
annually exceeds the quantity melted.*

125. Had we come in the month of April or May, we
should have found the glacier below the ice-fall also
covered with snow, which is now entirely cleared away
by the heat of summer. Nay, more, the ice there is
obviously melting, forming running brooks which cut
channels in the ice, and expand here and there into
small blue-green lakes. Hence you conclude with
certainty that below the ice-fall *the quantity of frozen
material falling upon the glacier is less than the quantity
melted.*

126. And this forces upon us another conclusion:

between the glacier below the ice-fall and the plateau above it there must exist a line where the quantity of snow which falls *is exactly equal* to the quantity annually melted. This is the *snow-line.* On some glaciers it is quite distinct, and it would be distinct here were the ice less broken and confused than it actually is.

127. The French term *névé* is applied to the glacial region above the snow-line, while the word *glacier* is restricted to the ice below it. Thus the snows of the Col du Géant constitute the névé of the Glacier du Géant, and in part, the néré of the Mer de Glace.

128. But if every year thus leaves a residue of snow upon the plateau of the Col du Géant, it necessarily follows that the plateau must get annually higher, *provided the snow remain upon it.* Equally certain is the conclusion that the whole length of the glacier below the cascade must sink gradually lower, *if the waste of annual melting be not made good.* Supposing two feet of snow a year to remain upon the Col, this would raise it to a height far surpassing that of Mont Blanc in five thousand years. Such accumulation must take place if the snow remain upon the Col; but the accumulation does *not* take place, hence the snow does not remain on the Col. The question then is, whither does it go?

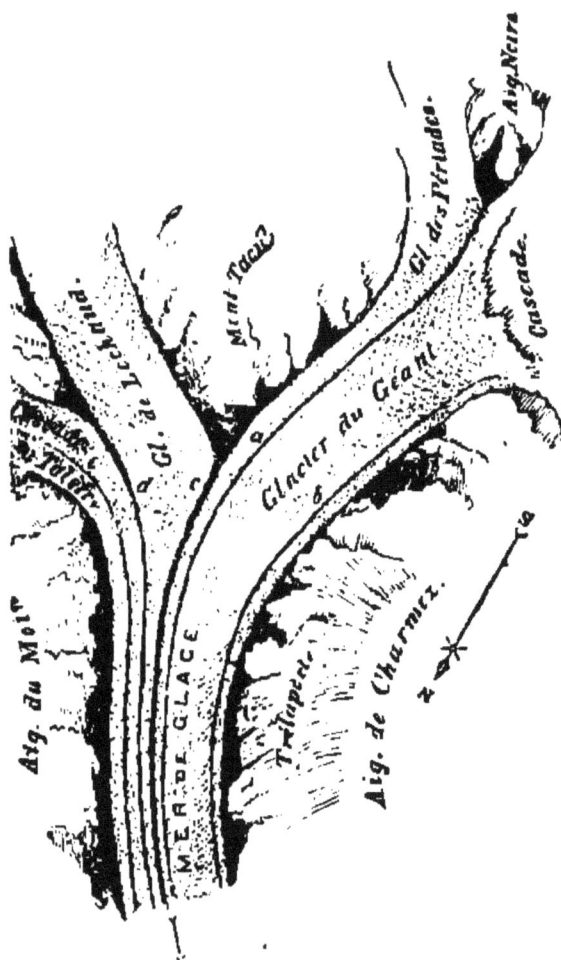

SKETCH-PLAN, SHOWING THE MORAINES, a, b, c, d, e, OF THE MER DE GLACE

§ 16. *Branches and Medial Moraines of the Mer de Glace from the Cleft Station.*

129. We shall grapple with this question immediately. Meanwhile look at that ice-valley in front of us, stretching up between Mont Tacul and the Aiguille de Léchaud, to the base of the great ridge called the Grande Jorasse. This is called the Glacier de Léchaud. It receives at its head the snows of the Jorasse and of Mont Mallet, and joins the Glacier du Géant at the promontory of the Tacul. The glaciers seem welded together where they join, but they continue distinct. Between them you clearly trace a stripe of débris (*c* on the annexed sketch-plan); you trace a similar though smaller stripe (*a* on the sketch), from the junction of the Glacier du Géant with the Glacier des Périades at the foot of the Aiguille Noire, which you also follow along the Mer de Glace.

130. We also see another glacier, or a portion of it, to the left, falling apparently in broken fragments through a narrow gorge (Cascade du Talèfre on the sketch) and joining the Léchaud, and from their point of junction also a stripe of débris (*d*) runs downwards along the Mer de Glace. Beyond this again we notice another stripe (*e*), which seems to begin at the bottom of the ice-fall, rising as it were from the body of the glacier. Beyond all of these we can notice the lateral moraine of the Mer de Glace.

■ 2

181. These stripes are the *medial moraines* of the Mer de Glace. We shall learn more about them immediately.

182. And now, having informed our minds by these observations, let our eyes wander over the whole glorious scene, the splintered peaks and the hacked and jagged crests, the far-stretching snow-fields, the smaller glaciers which nestle on the heights, the deep blue heaven and the sailing clouds. Is it not worth some labour to gain command of such a scene? But the delight it imparts is heightened by the fact that we did not come expressly to see it; we came to instruct ourselves about the glacier, and this high enjoyment is an incident of our labour. You will find it thus through life; without honest labour there can be no deep joy.

§ 17. *The Talèfre and the Jardin. Work among the Crevasses.*

133. And now let us descend to the Mer de Glace, for I want to take you across the glacier to that broken ice-fall the origin of which we have not yet seen. We aim at the further side of the glacier, and to reach it we must cross those dark stripes of débris which we observed from the heights. Looked at from above, these moraines seemed flat, but now we find them to be ridges of stones and rubbish, from twenty to thirty feet high.

134. We quit the ice at a place called the Couvercle, and wind round this promontory, ascending all the time. We squeeze ourselves through the *Égralets*, a kind of natural staircase in the rock, and soon afterwards obtain a full view of the ice-fall, the origin of which we wish to find. The ice upon the fall is much broken; we have pinnacles and towers, some erect, some leaning, and some, if we are fortunate, falling like those upon the Glacier des Bois; and we have chasms from which issues a delicate blue light. With the ice-fall to our right we continue to ascend, until at length we command a view of a huge glacier basin, almost level, and on the middle of which stands a solitary island, entirely surrounded by ice. We stand at the edge of the *Glacier du Talèfre*, and connect it with the ice-fall we have passed. The glacier is bounded by rocky ridges, hacked and torn at the top into teeth and edges, and buttressed by snow fluted by the descending stones.

135. We cross the basin to the central island, and find grass and flowers at the place where we enter upon it. This is the celebrated *Jardin*, of which you have often heard. The upper part of the Jardin is bare rock. Close at hand is one of the noblest peaks in this portion of the Alps, the Aiguille Verte. It is between thirteen and fourteen thousand feet high, and down its sides, after freshly-fallen snow, avalanches incessantly thunder. From one of its projections a streak of morino starts down the Talèfre; from the Jardin also a similar streak

of moraine issues. Both continue side by side to the top of the ice-fall, where they are engulphed in the chasms. But at the bottom of the fall they reappear, as if newly emerging from the body of the glacier, and afterwards they continue along the Mer de Glace.

136. Walk with me now alongside the moraine from the Jardin down towards the ice-fall. For a time our work is easy, such fissures as appear offering no impediment to our march. But the crevasses become gradually wider and wilder, following each other at length so rapidly as to leave merely walls of ice between them. Here perfect steadiness of foot is necessary—a slip would be death. We look towards the fall, and observe the confusion of walls and blocks and chasms below us increasing. At length prudence and reason cry 'Halt!' We may swerve to the right or to the left, and making our way along crests of ice, with chasms on both hands, reach either the right lateral moraine or the left lateral moraine of the glacier.

§ 18. *First Questions regarding Glacier Motion. Drifting of Bodies buried in a Crevasse.*

137. But what are these lateral moraines? As you and I go from day to day along the glaciers, their origin is gradually made plain. We see at intervals the stones and rubbish descending from the mountain sides and arrested by the ice. All along the fringe of the glacier the stones and rubbish fall, and it soon

becomes evident that we have here the source of the lateral moraines.

138. But how are the medial moraines to be accounted for? How does the débris range itself upon the glacier in stripes some hundreds of yards from its edge, leaving the space between them and the edge clear of rubbish? Some have supposed the stones to have rolled over the glacier from the sides, but the supposition will not bear examination. Call to mind now our reasoning regarding the excess of snow which falls above the snow-line, and our subsequent question, How is the snow disposed of. Can it be that the entire mass is moving slowly downwards? If so, the lateral moraines would be carried along by the ice on which they rest, and when two branch glaciers unite they would lay their adjacent lateral moraines together to form a medial moraine upon the trunk glacier.

139. There is, in fact, no way that we can see of disposing of the excess of snow above the snow-line; there is no way of making good the constant waste of the ice below the snow-line; there is no way of accounting for the medial moraines of the glacier, but by supposing that from the highest snow-fields of the Col du Géant, the Léchaud, and the Talèfre, to the extreme end of the Glacier des Bois, the whole mass of frozen matter is moving downwards.

140. If you were older, it would give me pleasure to take you up Mont Blanc. Starting from Chamouni, we

should first pass through woods and pastures, then up
the steep hill-face with the Glacier des Bossons to our
right, to a rock known as the *Pierre Pointue*; thence
to a higher rock called the *Pierre l'Échelle*, because here
a ladder is usually placed to assist in crossing the
chasms of the glacier. At the Pierre l'Échelle we
should strike the ice, and passing under the Aiguille du
Midi, which towers to the left, and which sometimes
sweeps a portion of the track with stone avalanches, we
should cross the Glacier des Bossons ; amid heaped-up
mounds and broken towers of ice ; up steep slopes ; over
chasms so deep that their bottoms are hid in darkness.

141. We reach the rocks of the Grands Mulets, which
form a kind of barren islet in the icy sea; thence
to the higher snow-fields, crossing the *Petit Plateau*,
which we should find cumbered by blocks of ice.
Looking to the right, we should see whence they came,
for rising here with threatening aspect high above us
are the broken ice-crags * of the Dôme du Goûté. The
guides wish to pass this place in silence, and it is just
as well to humour them, however much you may doubt
the competence of the human voice to bring the ice-
crags down. From the Petit Plateau a steep snow-slope
would carry us to the Grand Plateau, and at day-dawn
1 know nothing in the whole Alps more grand and
solemn than this place.

* Named *sérac* from their resemblance in shape and colour to an inferior
kind of curdy cheese called by this name at Chamouni.

· 142. One object of our ascent would be now attained; for here at the head of the Grand Plateau, and at the foot of the final slope of Mont Blanc, I should show you a great crevasse, into which three guides were poured by an avalanche in the year 1820.

143. Is this language correct? A crevasse hardly

CREVASSE ON GRAND PLATEAU.

to be distinguished from the present one undoubtedly · existed here in 1820. But was it the identical crevasse now existing? Is the ice riven here to-day the same as that riven fifty-one years ago? By no means. How is this proved? By the fact that more than forty years after their interment, the remains of those three guides

were found near the end of the Glacier des Bossons, many
miles below the existing crevasse.

144. The same observation proves to demonstration
that it is the ice near the *bottom* of the higher névé
that becomes the *surface-ice* of the glacier near its end.
The waste of the surface below the snow-line brings
the deeper portions of the ice more and more to the
light of day.

145. There are numerous obvious indications of the
existence of glacier motion, though it is too slow to
catch the eye at once. The crevasses change within
certain limits from year to year, and sometimes from
month to month; and this could not be if the ice did
not move. Rocks and stones also are observed, which
have been plainly torn from the mountain sides. Blocks
seen to fall from particular points are afterwards
observed lower down. On the moraines rocks are
found of a totally different mineralogical character from
those composing the mountains right and left; and in
all such cases strata of the same character are found
bordering the glacier higher up. Hence the conclusion
that the foreign boulders have been *floated* down by
the ice. Further, the ends or 'snouts' of many glaciers
act like ploughshares on the land in front of them,
overturning with slow but merciless energy huts and
châlets that stand in their way. Facts like these
have been long known to the inhabitants of the High

Alps, who were thus made acquainted in a vague and general way with the motion of the glaciers.

§ 19. *The Motion of Glaciers. Measurements by Hugi and Agassiz. Drifting of Huts on the Ice.*

146. But the growth of knowledge is from vagueness towards precision, and exact determinations of the rate of glacier motion were soon desired. With reference to such measurements one glacier in the Bernese Oberland will remain for ever memorable. From the little town of Meyringen in Switzerland you proceed up the valley of Hasli, past the celebrated waterfall of Handeck, where the river Aar plunges into a chasm more than 200 feet deep. You approach the Grimsel Pass, but instead of crossing it you turn to the right and follow the course of the Aar upwards. Like the Rhone and the Arveiron, you find the Aar issuing from a glacier.

147. Get upon the ice, or rather upon the deep moraine shingle which covers the ice, and walk upwards. It is hard walking, but after some time you get clear of the rubbish, and on to a wide glacier with a great medial moraine running along its back. This moraine is formed by the junction of two branch glaciers, the Lauteraar and the Finsteraar, which unite at a promontory called the Abschwung to form the trunk glacier of the Unteraar.

148. On this great medial moraine in 1827 an intrepid

and enthusiastic Swiss professor, Hugi, of Solothurm (French Soleure), built a hut with a view to observations upon the glacier. His hut moved, and he measured its motion. In the three years—from 1827 to 1830—it had moved 330 feet downwards. In 1836 it had moved 2,354 feet; and in 1841 M. Agassiz found it 4,712 feet below its first position.

149. In 1840, M. Agassiz himself and some bold companions took shelter under a great overhanging slab of rock on the same moraine, to which they added side walls and other means of protection. And because he and his comrades came from Neufchatel, the hut was called long afterwards the ' Hôtel des Neuchâtelois.' Two years subsequent to its erection M. Agassiz found that the ' hotel ' had moved 486 feet downwards.

§ 20. *Precise Measurements of Agassiz and Forbes. Motion of a Glacier proved to resemble the Motion of a River.*

150. We now approach an epoch in the scientific history of glaciers. Had the first observers been practically acquainted with the instruments of precision used in surveying, *accurate* measurements of the motion of glaciers would probably have been earlier executed. We are now on the point of seeing such instruments introduced almost simultaneously by M. Agassiz on the glacier of the Unteraar, and by Professor Forbes on the Mer de Glace. Attempts had been made

by M. Escher de la Linth to determine the motion of a series of wooden stakes driven into the Aletsch glacier, but the melting was so rapid that the stakes soon fell. To remedy this, M. Agassiz in 1841 undertook the great labour of carrying boring tools to his 'hotel,' and piercing the Unteraar glacier at six different places to a depth of ten feet, in a straight line across the glacier. Into the holes six piles were so firmly driven that they remained in the glacier for a year, and in 1842 the displacements of all six were determined. They were found to be 160 feet, 225 feet, 269 feet, 245 feet, 210 feet, and 125 feet, respectively.

151. A great step is here gained. You notice that the middle numbers are the largest. They correspond to the central portion of the glacier. Hence, these measurements conclusively establish, not only the fact of glacier motion, but that *the centre of the glacier, like that of a river, moves more rapidly than the sides.*

152. With the aid of trained engineers M. Agassiz followed up these measurements in subsequent years. His researches are recorded in a work entitled 'Système glaciaire,' which is accompanied by a very noble Atlas of the Glacier of the Unteraar, published in 1847.

153. These determinations were made by means of a theodolite, of which I will give you some notion immediately. The same instrument was employed the same year by the late Principal Forbes upon the Mer de Glace. He established independently the greater central

motion. He showed, moreover, that it is not necessary
to wait a year, or even a week to determine the motion
of a glacier; with a correctly-adjusted theodolite he
was able to determine the motion of various points of
the Mer de Glace from day to day. He affirmed, and
with truth, that the motion of the glacier might be
determined from hour to hour. We shall prove this
further on (162). Professor Forbes also triangulated
the Mer de Glace, and laid down an excellent map of
it. His first observations and his survey are recorded
in a celebrated book published in 1843, and entitled
'Travels in the Alps.'

154. These observations were also followed up in
subsequent years, the results being recorded in a series
of detached letters and essays of great interest. These
were subsequently collected in a volume entitled 'Occa-
sional Papers on the Theory of Glaciers,' published in
1859. The labours of Agassiz and Forbes are the two
chief sources of our knowledge of glacier phenomena.

§ 21. The Theodolite and its Use. Our own Measurements.

155. My object thus far is attained. I have given
you proofs of glacier motion, and a historic account of
its measurement. And now we must try to add a
little to the knowledge of glaciers by our own labours
on the ice. Resolution must not be wanting at the
commencement of our work, nor steadfast patience

during its prosecution. Look then at this theodolite; it consists mainly of a telescope and a graduated circle, the telescope capable of motion up and down, and the circle, carrying the telescope along with it, capable of motion right and left. When desired to make the motion exceedingly fine and minute, suitable screws, called tangent screws, are employed. The instrument is supported by three legs, movable, but firm when properly planted.

156. Two spirit-levels are fixed at right angles to each other on the circle just referred to. Practice enables one to take hold of the legs of the instrument. and so to fix them that the circle shall be nearly horizontal. By means of four levelling screws we render it *accurately* horizontal. Exactly under the centre of the instrument is a small hook from which a plummet is suspended; the point of the bob just touches a rock on which we make a mark; or if the earth be soft underneath, we drive a stake into it exactly under the plummet. By re-suspending the plummet at any future time we can find to a hairbreadth the position occupied by the instrument to-day.

157. Look through the telescope; you see it crossed by two fibres of the finest spider's thread. In actual work we first direct the telescope across the glacier, until the intersection of the two fibres accurately covers some well-defined point of rock or tree at the other side of the valley. This, our fixed standard, we sketch

with its surroundings in a note-book, so as to be able
immediately to recognise it on our return to this place.
Imagine a straight line drawn from the centre of the
telescope to this point, and that this line is permitted
to drop straight down upon the glacier, every point of
it falling as a stone would fall; along such a line we
have now to fix a series of stakes.

158. A trained assistant is already upon the glacier.
He erects his staff and stands behind it; the telescope
is lowered without swerving to the right or to the
left; in mathematical language it remains *in the same
vertical plane.* The crossed fibres of the telescope
probably strike the ice a little away from the staff of
the assistant; by a wave of the arm he moves right or
left; he may move too much, so we wave him back
again. After a trial or two he knows whether he is
near the proper point, and if so makes his motions
small. He soon exactly strikes the point covered by
the intersection of the fibres. A signal is made which
tells him that he is right; he pierces the ice with an
auger and drives in a stake. He then goes forward, and
in precisely the same manner takes up another point.
After one or two stakes have been driven in, the
assistant is able to take up the other points very
rapidly. Any requisite number of stakes may thus be
fixed in a straight line across the glacier.

159. Next morning we measure the motion of all the
stakes. The theodolite is mounted in its former posi-

tion and carefully levelled. The telescope is directed
first upon the standard point at the opposite side of
the valley, being moved by a tangent screw until the
intersection of the spider's threads accurately covers
the point. The telescope is then lowered to the
first stake, beside which our trained assistant is
already standing. He is provided with a staff with
feet and inches marked on it. A glance shows us
that the stake has moved down. By our signals the
assistant recovers the point from which we started
yesterday, and then determines the distance from this
point to the stake. It is, say, 6 inches; through
this distance, therefore, the stake has moved.

160. We are careful to note the hour and minute at
which each stake is driven in, and the hour and the
minute when its distance from its first position is
measured; this enables us to calculate the accurate
daily motion of the point in question. The distances
through which all the other points have moved are
determined in precisely the same way.

161. Thus we shall proceed to work, first making
clear to our minds what is to be done, and then making
sure that it shall be accurately done. To give our work
reality, I will here record the actual measurements
executed, and the actual thoughts suggested, on the
Mer de Glace in 1857. The only unreality that I
would ask you to allow, is that you and I are supposed
to be making the observations together. The labour

P

of measuring was undertaken for the most part by Mr. Hirst.

§ 22. *Motion of the Mer de Glace.*

162. On July 14, then, we find ourselves at the end of the Glacier des Bois, not far from the source of the Arveiron. We direct our telescope across the glacier, and fix the intersection of its spider's threads accurately upon the edge of a pinnacle of ice. We leave the instrument untouched, looking through it from hour to hour. The edge of ice moves slowly, but plainly, past the fibres, and at the end of three hours we assure ourselves that the motion has amounted to several inches. While standing near the vault of the Arveiron, and talking about going into it, its roof gives way, and falls with the sound of thunder. It is not, therefore, without reason that I warned you against entering these vaults in summer.

163. We ascend to the Montanvert Inn, fix on it as a residence, and then descend to the lateral moraine of the glacier a little below the inn. Here we erect our theodolite, and mark its exact position by a plummet. We must first make sure that our line is perpendicular, or nearly so, to the axis or middle line of the glacier. Our instructed assistant lays down a long staff in the direction of the axis, assuring himself, by looking up and down, that it is the true direction. With another staff in his hand, pointed towards our theodolite, he shifts his position until the second staff is perpendicular to the

first. Here he gives us a signal. We direct our telescope upon him, and then gradually raising its end in a
vertical plane we find, and note by sketching, a standard
point at the other side of the glacier. This point known,
and our plummet mark known, we can on any future
day find our line. (To render the measurements more
intelligible, I append on the next page an outline diagram of the Mer de Glace, and of its tributaries.)

164. Along the line just described ten stakes were set
on July 17, 1857. Their displacements were measured
on the following day. Two of them had fallen, but
here are the distances passed over by the eight remaining ones in twenty-four hours.

DAILY MOTION OF THE MER DE GLACE.

First Line: A A' upon the Sketch.

	East							West
Stake . . .	1	2	3	4	5	7	9	10
Inches . . .	12	17	23	26	25	26	27	33

165. You have already assured yourself by actual
contact that the body of the glacier is real ice, and you
may have read that glaciers move; but the actual
observation of the motion of a body apparently so rigid
is strangely interesting. And not only does the ice
move bodily, but one part of it moves past another;
the rate of motion augmenting gradually from 12
inches a day at the side to 33 inches a day at a
distance from the side. This quicker movement of the
central ice of glaciers had been already observed by

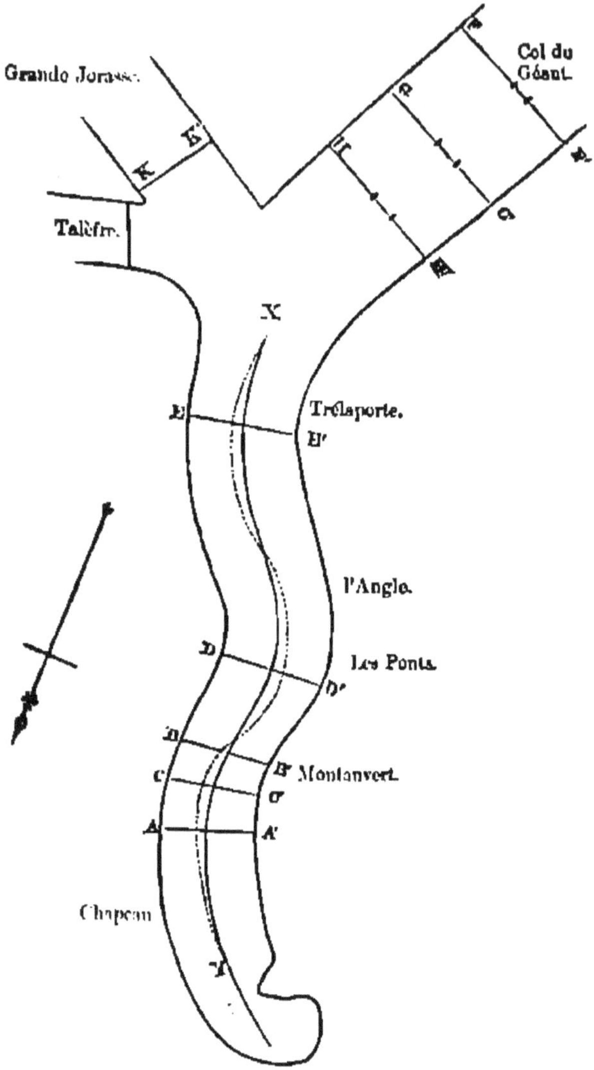

Grande Jorass.

Col du
Géant

K

Taléfr.

X

E
Trélaporte.
H'

l'Angle.

D
Les Ponts.
D'

B
C
Montanvert.
C'

A
A'

Chapeau

OUTLINE PLAN, SHOWING THE MEASURED LINES OF THE MER DE GLACE AND
ITS TRIBUTARIES.

Agassiz and Forbes; we verify their results, and now proceed to something new. Crossing the Glacier du Géant, which occupies more than half the valley, we find that our line of stakes is not yet at an end. The 10th stake stands on the part of the ice which comes from the Talèfre.

166. Now the motion of the sides is slow, because of the friction of the ice against its boundaries; but then one would think that midway between the boundaries, where the friction of the sides is least, the motion ought to be greatest. This is clearly not the case; for though the 10th stake is nearer than the 9th to the eastern or *Chapeau* side of the valley, the 10th stake surpasses the 9th by 6 inches a day.

167. Here we have something to think of; but before a natural philosopher can think with comfort he must be perfectly sure of his facts. The foregoing line ran across the glacier a little below the Montanvert. We will run another line across a little way above the hotel. On July 18 we set out this line, and to multiply our chances of discovery we place along it 31 stakes. On the subsequent day five of these were found unfit for use; but here are the distances passed over by the remaining six-and-twenty in 24 hours.

SECOND LINE: B F upon the Sketch.

West												
Stake	2	3	4	5	6	7	8	9	10	11	12	13
Inches	11	12	15	15	16	17	18	19	20	20	21	21
Stake	15	16	17	18	19	20	21	22	23	24	25	26
Inches	23	23	23	21	23	21	25	22	22	23	25	26

East

168. Look at these numbers. The first broad fact they reveal is the advance in the rate of motion from first to last. There are however small irregularities; from 23 inches at the 17th stake we fall to 21 inches at the 18th; from 23 inches at the 19th we fall to 21 inches at the 20th; from 25 inches at the 21st we fall to 22 inches at the 22nd and 23rd; but notwithstanding these small ups and downs, the general advance of the rate of motion is manifest. Now there may have been some slight displacement of the stakes by melting, sufficient to account for these small deviations from uniformity in the increase of the motion. But another solution is also possible. We shall afterwards learn that the glacier is retarded not only by its sides but by its bed; that the upper portions of the ice slide over the lower ones. Now if the bed of the Mer de Glace should have eminences here and there rising sufficiently near to the surface to retard the motion of the surface, they might produce the small irregularities noticed above.

169. We note particularly, while upon the ice, that the 26th stake, like the 10th stake in our last line, stands much nearer to the eastern than to the western side of the glacier; the measurements, therefore, offer a further proof that the centre of this portion of the glacier is *not* the place of swiftest motion.

§ 23. *Unequal Motion of the two Sides of the
Mer de Glace.*

170. But in neither the first line nor the second were

we able to push our measurements quite across the
glacier. Why? In attempting to do one thing we are
often taught another, and thus in science, if we are only
steadfast in our work, our very defeats are converted
into means of instruction. We at first planted our
theodolite on the lateral moraine of the Mer de Glace,
expecting to be able to command the glacier from side
to side. But we are now undeceived; the centre of
the glacier proves to be higher than its sides, and from
our last two positions the view of the ice near the opposite side of the glacier was intercepted by the elevation at the centre. The mountain slopes, in fact, are
warm in summer, and they melt the ice nearest to them,
thus causing a fall from the centre to the sides.

171. But yonder on the heights at the other side of
the glacier we see a likely place for our theodolite.
We cross the glacier and plant our instrument in a
position from which we sweep the glacier from side to
side. Our first line was below the Montanvert, our
second line above it; this third line is exactly opposite
the Montanvert; in fact, the mark on which we have
fixed the fibre-cross of the theodolite is a corner of one
of the windows of the little inn. Along this line we
fix twelve stakes on July 20. On the 21st one of them
had fallen; but the velocities of the remaining eleven
in 24 hours were found to be as follows :—

THIRD LINE: CC' UPON THE SKETCH.

	East									West	
Stake . .	1	2	3	4	5	6	7	8	9	10	11
Inches .	20	23	29	30	34	28	25	25	25	18	9

172. Both the first stake and the eleventh in this series stood near the sides of the glacier. On the eastern side the motion is 20 inches, while on the western side it is only 9. It rises on the eastern side from 20 to 34 inches at the 5th stake, which we, standing upon the glacier, can see to be much nearer to the eastern than to the western side. *The united evidence of these three lines places the fact beyond doubt, that opposite the Montanvert, and for some distance above it and below it, the whole eastern side of the glacier is moving more quickly than the western side.*

§ 24. *Suggestion of a new Likeness of Glacier Motion to River Motion. Conjecture tested.*

173. Here we have cause for reflection, and facts are comparatively worthless if they do not provoke this exercise of the mind. It is because facts of nature are not isolated but connected, that science, to follow them, must also form a connected whole. The mind of the natural philosopher must, as it were, be a web of *thought* corresponding in all its fibres with the web of *fact* in nature.

174. Let us, then, ascend to a point which commands a good view of this portion of the Mer de Glace. The

ice-river we see is not straight but curved, and its curvature is *from* the Montanvert; that is to say, its convex side is east, and its concave side is west (look to
the sketch). You have already pondered the fact that
a glacier, *in some respects,* moves like a river. How would
a river move through a curved channel? This is known.
Were the ice of the Mer de Glace displaced by water,
the point of swiftest motion at the Montanvert would
not be the centre, but a point east of the centre. Can
it be then that this 'water-rock,' as ice is sometimes
called, acts in this respect also like water?

175. This is a thought suggested on the spot; it may
or it may not be true, but the means of testing it are at
hand. Looking up the glacier, we see that at *les Ponts*
it also bends, but that there its convex curvature is
towards the western side of the valley (look again to
the sketch). If our surmise be true, the point of
swiftest motion opposite *les Ponts* ought to lie west of
the axis of the glacier.

176. Let us test this conjecture. On July 25 we fix
in a line across this portion of the glacier seventeen
stakes; every one of them has remained firm, and on
the 26th we find the motion for 24 hours to be as
follows:—

FOURTH LINE: DD′ UPON THE SKETCH.

East															West
Stake . 1	2	3	4	5	6	7	8	9	10	11	12	13	14	15
Inches . 7	8	13	15	16	19	20	21	21	23	23	21	22	17	15

177. Inspected by the naked eye alone, the stakes 10 and 11, where the glacier reaches its greatest motion, are seen to be considerably to the west of the axis of the glacier. Thus far we have a perfect verification of the *guess* which prompted us to make these measurements. You will here observe that the 'guesses' of science are not the work of chance, but of thoughtful pondering over antecedent facts. The guess is the 'induction' from the facts, to be ratified or exploded by the test of subsequent experiment.

178. And though even now we have exceedingly strong reason for holding that the point of maximum velocity obeys the law of liquid motion, the strength of our conclusion will be doubled if we can show that the point shifts back to the eastern side of the axis at another place of flexure. Fortunately such a place exists opposite Trélaporte. Here the convex curvature of the valley turns again to the east. Across this portion of the glacier a line was set out on July 28, and from measurements on the 31st, the rate of motion per 24 hours was determined.

FIFTH LINE: E E' UPON THE SKETCH.

West														East
Stake . 1	2	3	4	5	6	7	8	9	10	11	12	13	14	15
Inches . 11	14	13	15	15	16	17	19	20	19	20	18	16	15	10

179. Here, again, the mere estimate of distances by the eye would show us that the three stakes which moved fastest, viz. the 9th, 10th, and 11th, were all to the east of

the middle line of the glacier. The demonstration that
the point of swiftest motion wanders to and fro across
the axis, as the flexure of the valley changes, is, therefore,—shall I say complete?

180. Not yet. For if surer means are open to us we
must not rest content with estimates by the eye. We
have with us a surveying chain: let us shake it out and
measure these lines, noting the distance of every stake
from the side of the glacier. This is no easy work
among the crevasses, but I confide it confidently to Mr.
Hirst and you. We can afterwards compare a number
of stakes on the eastern side with the same number of
stakes taken at the same distances from the western
side. For example, a pair of stakes, one ten yards from
the eastern side and the other ten yards from the
western side; another pair, one fifty yards from the
eastern side and the other fifty yards from the western
side, and so on, can be compared together. For the
sake of easy reference, let us call the points thus compared in pairs, *equivalent points.*

181. There were five pairs of such points upon our
fourth line, D D', and here are their velocities:—

| Eastern points; motion in inches | . | . | 13 | 15 | 16 | 18 | 20 |
| Western points | „ | „ | . | . | 16 | 17 | 22 | 23 | 23 |

In every case here the stake at the western side moved
more rapidly than the equivalent stake at the eastern
side.

182. Applying the same analysis to our fifth line,

E E', we have the following series of velocities of three pairs of equivalent points :—

Eastern points; motion in inches . . 15   18   19
Western points   „   „   . . 13   15   17

183. Here the three points on the eastern side move more rapidly than the equivalent points on the western side.

184. It is thus proved :—

1. *That opposite the Montanvert the eastern half of the Mer de Glace moves more rapidly than the western half.*

2. *That opposite* les Ponts *the western half of the glacier moves more rapidly than the eastern half.*

3. *That opposite Trélaporte the eastern half of the glacier again moves more rapidly than the western half.*

4. *That these changes in the place of greatest motion are determined by the flexures of the valley through which the Mer de Glace moves.*

§ 25. *New Law of Glacier Motion.*

185. Let us express these facts in another way. Supposing the points of swiftest motion for a very great number of lines crossing the Mer de Glace to be determined; the line joining all those points together is what mathematicians would call the *locus* of the point of swiftest motion.

186. At Trélaporte this line would lie east of the centre; at the *Ponts* it would lie west of the centre; hence in passing from Trélaporte to the *Ponts* it would

cross the centre. But at the Montanvert it would again
lie east of the centre; hence between the *Ponts* and the
Montanvert the centre must be crossed a second time.
If there were further sinuosities upon the Mer de Glace
there would be further crossings of the axis of the glacier.

187. The points on the axis which mark the transition
from eastern to western bending, and the reverse, may
be called *points of contrary flexure.*

188. Now what is true of the Mer de Glace is true of
all other glaciers moving through sinuous valleys; so
that the facts established in the Mer de Glace may be
expanded into the following general law of glacier motion :—

*When a glacier moves through a sinuous valley, the locus
of the point of maximum motion does not coincide with
the centre of the glacier, but, on the contrary, always
lies on the convex side of the central line. The locus is
therefore a curved line more deeply sinuous than the valley
itself, and crosses the axis of the glacier at each point of
contrary flexure.*

189. The dotted line on the Outline Plan (page 68)
represents the locus of the point of maximum motion,
the firm line marking the centre of the glacier.

190. Substituting the word *river* for *glacier*, this law
is also true. The motion of the water is ruled by precisely the same conditions as the motion of the ice.

191. Let us now apply our law to the explanation
of a difficulty. Turning to the careful measurements

executed by M. Agassiz on the glacier of the Unteraar, we notice in the discussion of these measurements a section of the 'Système glaciaire' devoted to the 'Migrations of the Centre.' It is here shown that the middle of the Unteraar glacier is not always the point of swiftest motion. This fact has hitherto remained without explanation ; but a glance at the Unteraar valley, or at the map of the valley, shows the enigma to be an illustration of the law which we have just established on the Mer de Glace.

§ 26. *Motion of Axis of Mer de Glace.*

192. We have now measured the rate of motion of five different lines across the trunk of the Mer de Glace. Do they all move alike ? No. Like a river, a glacier at different places moves at different rates. Comparing together the points of maximum motion of all five lines, we have this result :—

MOTION OF MER DE GLACE.

At Trélaporte	.	.	. 20 inches a day.	
At *les Ponts*	.	.	. 23	„ „
Above the Montanvert	.	. 26	„	„
At the Montanvert	.	. 34	„	„
Below the Montanvert	.	. 33*	„	„

193. There is thus an increase of rapidity as we descend the glacier from Trélaporte to the Montanvert ;

---

* This is probably under the mark. I think it likely that the swiftest motion of this portion of the Mer de Glace in 1857 amounted to a yard in twenty-four hours.

the maximum motion at the Montanvert being fourteen inches a day greater than at Trélaporte.

§ 27. *Motion of Tributary Glaciers.*

194. So much for the trunk glacier; let us now investigate the branches, permitting, as we have hitherto done, reflection on known facts to precede our attempts to discover unknown ones.

195. As we stood upon our 'cleft station,' whence we had so capital a view of the Mer de Glace, we were struck by the fact that some of the tributaries of the glacier were wider than the glacier itself. Supposing water to be substituted for the ice, how do you suppose it would behave? You would doubtless conclude that the motion down the broad and slightly-inclined valleys of the Géant and the Léchaud would be comparatively slow, but that the water would force itself with increased rapidity through the 'narrows' of Trélaporte. Let us test this notion as applied to the ice.

196. Planting our theodolite in the shadow of Mont Tacul, and choosing a suitable point at the opposite side of the Glacier du Géant, we fix on July 29 a series of ten stakes across the glacier. The motion of this line in twenty-four hours was as follows :—

MOTION OF GLACIER DU GÉANT.

SIXTH LINE: H H' UPON SKETCH.

Stake	1	2	3	4	5	6	7	8	9	10
Inches	11	10	13	13	12	13	11	10	9	6

197. Our conjecture is fully verified. The maximum motion here is seven inches a day less than that of the Mer de Glace at Trélaporte (192).

198. And now for the Léchaud branch. On August 1 we fix ten stakes across this glacier above the point where it is joined by the Talèfre. Measured on August 3, and reduced to twenty-four hours, the motion was found to be—

MOTION OF GLACIER DE LÉCHAUD.

SEVENTH LINE: K K' UPON SKETCH.

Stake . .	1	2	3	4	5	6	7	8	9	10
Inches .	5	8	10	9	9	8	6	9	7	6

199. Here our conjecture is still further verified, the rate of motion being even less than that of the Glacier du Géant.

§ 28. *Motion of Top and Bottom of Glacier.*

200. We have here the most ample and varied evidence that the sides of a glacier, like those of a river, are retarded by friction against its boundaries. But the likeness does not end here. The motion of a river is retarded by the friction against its bed. Two observers, viz. Prof. Forbes and M. Charles Martins, concur in showing the same to be the case with a glacier. The observations of both have been objected to; hence it is all the more incumbent on us to seek for decisive evidence.

201. At the Tacul (near the point a upon the sketch

plan, p. 83) a wall of ice about 150 feet high has already attracted our attention. Bending round to join the Léchaud the Glacier du Géant is here drawn away from the mountain side, and exposes a fine section. We try to measure it top, bottom, and middle, and are defeated twice over. We try it a third time and succeed. A stake is fixed at the summit of the ice-precipice, another at 4 feet from the bottom, and a third at 35 feet above the bottom. These lower stakes are fixed at some risk of boulders falling upon us from above; but by skill and caution we succeed in measuring the motions of all three. For 24 hours the motions are :—

Top stake	. . . .	6 inches.
Middle stake	. . . .	4½ "
Bottom stake	. . . .	2⅔ "

202. The retarding influence of the bed of the glacier is reduced to demonstration by these measurements. The bottom does not move with half the velocity of the surface.

§ 29. *Lateral Compression of a Glacier.*

203. Furnished with the knowledge which these labours and measurements have given us, let us once more climb to our station beside the Cleft under the Aiguille de Charmoz. At our first visit we saw the medial moraines of the glacier, but we knew nothing about their cause. We now know that they mark upon the trunk its tributary glaciers. Cast your eye, then,

first upon the Glacier du Géant; realise its width in
its own valley, and see how much it is narrowed at
Trélaporte.   The broad ice-stream of the Léchaud is
still more surprising, being squeezed upon the Mer de
Glace to a narrow white band between its bounding
moraines.   The Talèfre undergoes similar compression.
Let us now descend, shake out our chain, measure,
and express in numbers the width of the tributaries,
and the actual amount of compression suffered at Trélaporte.

204. We find the width of the Glacier du Géant to
be 5,155 links, or 1,134 yards.

205. The width of the Glacier de Léchaud we find
to be 3,725 links, or 825 yards.

206. The width of the Talèfre we find to be 2,900
links, or 638 yards.

207. The sum of the widths of the three branch
glaciers is therefore 2,597 yards.

208. At Trélaporte these three branches are forced
through a gorge 893 yards wide, or one-third of their
previous width, at the rate of twenty inches a day.

209. If we limit our view to the Glacier de Léchaud,
the facts are still more astonishing.   Previous to its
junction with the Talèfre, this glacier has a width of
825 yards; in passing through the jaws of the granite
vice at Trélaporte, its width is reduced to eighty-eight
yards, or in round numbers to one-tenth of its previous
width.   (Look to the sketch on the next page.)

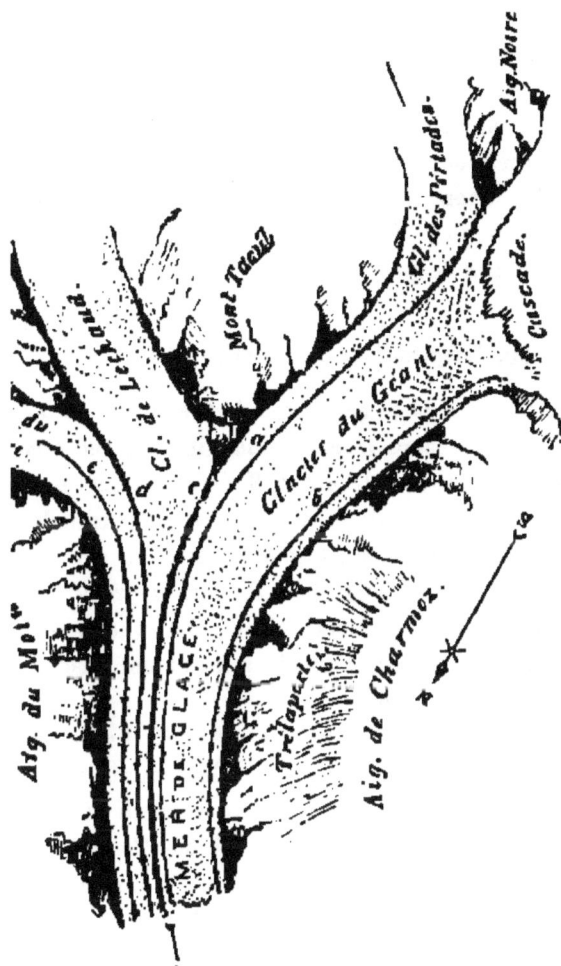

SKETCH-PLAN SHOWING THE MORAINES a, b, c, d, e, OF THE MER DE GLACE.

a 2

210. Are we to understand by this that the ice of the Léchaud is squeezed to one-tenth of its former *volume?* By no means. It is mainly a change of *form*, not of volume, that occurs at Trélaporte. Previous to its compression, the glacier resembles a plate of ice *lying flat* upon its bed. After its compression, it resembles a plate *fixed upon its edge.* The squeezing, doubtless, has deepened the ice.

§ 30. *Longitudinal Compression of a Glacier.*

211. The ice is forced through the gorge at Trélaporte by a pressure from behind; in fact the Glacier du Géant, immediately above Trélaporte, represents a piston or a plug which drives the ice through the gorge. What effect must this pressure have upon the plug itself? Reasoning alone renders it probable that the pressure will shorten the plug; that the lower part of the Glacier du Géant will to some extent yield to the pressure from behind.

212. Let us test this notion. About three-quarters of a mile above the Tacul, and on the mountain slope to the left as we ascend, we observe a patch of verdure. Thither we climb; there we plant our theodolite, and set out across the Glacier du Géant, a line, which we will call line No. 1 (F F′ upon sketch, p. 68).

213. About a quarter of a mile lower down we find a practicable couloir on the mountain side; we ascend

it, reach a suitable platform, plant our instrument, and
set out a second line, No. 2 (G G' upon sketch). We
must hasten our work here, for along this couloir stones
are discharged from a small glacier which rests upon
the slope of Mont Tacul.

214. Still lower down by another quarter of a mile,
which brings us near the Tacul, we set out a third line,
No. 3 (H H' upon sketch), across the glacier.

215. The daily motion of the centres of these three
lines is as follows:—

	Inches	Distances asunder
No. 1 .	. 20·55	. 545 yards.
No. 2 .	. 15·43	
No. 3 .	. 12·75	. 487 ,,

216. The first line here moves five inches a day more
than the second ; and the second nearly three inches a
day more than the third. The reasoning is therefore
confirmed. The ice-plug, which is in round numbers
one thousand yards long, is shortened by the pressure
exerted on its front at the rate of about eight inches a
day.

217. A river descending the Valley du Géant would
behave in substantially the same fashion. It would have
its motion on approaching Trélaporte diminished, and
it would pour through the defile with a velocity greater
than that of the water behind.

§ 31. *Sliding and Flowing. Hard Ice and Soft Ice.*

218. We have thus far confined ourselves to the measurement and discussion of glacier motion; but in our excursions we have noticed many things besides. Here and there, where the ice has retreated from the mountain side, we have seen the rocks fluted, scored, and polished; thus proving that the ice had slidden over them and ground them down. At the source of the Arveiron we noticed the water rushing from beneath the glacier charged with fine matter. All glacier rivers are similarly charged. The Rhone carries its load of matter into the Lake of Geneva; the rush of the river is here arrested, the matter subsides, and the Rhone quits the lake clear and blue. The Lake of Geneva, and many other Swiss lakes, are in part filled up with this matter, and will, in all probability, finally be obliterated by it.

219. One portion of the motion of a glacier is due to this bodily sliding of the mass over its bed.

220. We have seen in our journeys over the glacier streams formed by the melting of the ice, and escaping through cracks and *crevasses* to the bed of the glacier. The fine matter ground down is thus washed away; the bed is kept lubricated, and the sliding of the ice rendered more easy than it would otherwise be.

221. As a skater also you know how much ice is weakened by a thaw. Before it actually melts it becomes

rotten and unsafe. Test such ice with your penknife:
you can dig the blade readily into it, or cut the ice with
ease. Try good sound ice in the same way: you find
it much more resistant. The one, indeed, resembles
soft chalk; the other hard stone.

222. Now the Mer de Glace in summer is in this
thawing condition. Its ice is rendered soft and yielding
by the sun; its motion is thereby facilitated. We
have seen that not only does the glacier slide over its
bed, but that the upper layers slide over the under
ones, and that the centre slides past the sides. The
softer and more yielding the ice is, the more free will
be this motion, and the more readily also will it be
forced through a defile like Trélaporte.

223. But in winter the thaw ceases; the quantity of
water reaching the bed of the glacier is diminished or
entirely cut off. The ice also, to a certain depth at
least, is frozen hard. These considerations would justify
the opinion that in winter the glacier, if it moves at
all, must move more slowly than in summer. At all
events, the summer measurements give no clue to the
winter motion.

224. This point merits examination. I will not, however, ask you to visit the Alps in mid-winter; but, if
you allow me, I will be your deputy to the mountains,
and report to you faithfully the aspect of the region
and the behaviour of the ice.

§ 32. *Winter on the Mer de Glace.*

225. The winter chosen is an inclement one. There is snow in London, snow in Paris, snow in Geneva; snow near Chamouni so deep that the road fences are entirely effaced. On Christmas night—nearly at midnight—1859, your deputy reaches Chamouni.

226. The snow fell heavily on December 26; but on the 27th, during a lull in the storm, we turn out. There are with me four good guides and a porter. They tie planks to their feet to prevent them from sinking in the snow; I neglect this precaution and sink often to the waist. Four or five times during our ascent the slope cracks with an explosive sound, and the snow threatens to come down in avalanches.[*]

The freshly-fallen snow was in that particular condition which causes its granules to adhere, and hence every flake falling on the trees had been retained there. The laden pines presented beautiful and often fantastic forms.

227. After five hours and a half of arduous work the Montanvert was attained. We unlocked the forsaken auberge, round which the snow was reared in buttresses. I have already spoken of the complex play of crystallising

---

[*] Four years later, viz. in the spring of 1863, a mighty climber and noble guide and companion of mine, named Johann Joseph Bennen, was lost, through the cracking and subsequent slipping of snow on such a slope.

forces. The frost-figures on the window-panes of the auberge were wonderful: mimic shrubs and ferns wrought by the building power while hampered by the adhesion between the glass and the film in which it worked. The appearance of the glacier was very impres-

SNOW-LADEN PINE-TREE.

sive; all sounds were stilled. The cascades which in summer fill the air with their music were silent, hanging from the ledges of the rocks in fluted columns of ice. The surface of the glacier was obviously higher than it had been in summer; suggesting the thought that while the winter cold maintained the lower end of the glacier

jammed between its boundaries, the upper portions still
moved downwards and thickened the ice. The peak of
the Aiguille du Dru shook out a cloud-banner, the origin and nature of which have been already explained (84).
(See *Frontispiece.*)

228. On the morning of the 28th this banner was
strikingly large and grand, and reddened by the light
of the rising sun, it glowed like a flame. Roses of
cloud also clustered round the crests of the Grande
Jorasse and hung upon the pinnacles of Charmoz.
Four men, well roped together, descended to the glacier.
I had trained one of them in 1857, and he was now to
fix the stakes. The storm had so distributed the snow
as to leave alternate lengths of the glacier bare and
thickly covered. Where much snow lay great caution
was required, for hidden crevasses were underneath.
The men sounded with their staffs at every step. Once
while looking at the party through my telescope the
leader suddenly disappeared; the roof of a crevasse had
given way beneath him; but the other three men
promptly gathered round and lifted him out of the
fissure. The true line was soon picked up by the theodolite; one by one the stakes were fixed until a series
of eleven of them stood across the glacier.

229. To get higher up the valley was impracticable;
the snow was too deep, and the aspect of the weather
too threatening; so the theodolite was planted amid the
pines a little way below the Montanvert, whence through

a vista I could see across the glacier. The men were wrapped at intervals by whirling snow-wreaths which quite hid them, and we had to take advantage of the lulls in the wind. Fitfully it came up the valley, darkening the air, catching the snow upon the glacier, and tossing it throughout its entire length into high and violently agitated clouds, separated from each other by cloudless spaces corresponding to the naked portions of the ice. In the midst of this turmoil the men continued to work. Bravely and steadfastly stake after stake was set, until at length a series of ten of them was fixed across the glacier.

230. Many of the stakes were fixed in the snow. They were four feet in length, and were driven in to a depth of about three feet. But that night, while listening to the wild onset of the storm, I thought it possible that the stakes and the snow which held them might be carried bodily away before the morning. The wind, however, lulled. We rose with the dawn, but the air was thick with descending snow. It was all composed of those exquisite six-petaled flowers, or six-rayed stars, which have been already figured and described (§ 9). The weather brightening, the theodolite was planted at the end of the first line. The men descended, and, trained by their previous experience, rapidly executed the measurements. The first line was completed before 11 A.M. Again the snow began to fall, filling all the air. Spangles innumerable were showered

upon the heights. Contrary to expectation, the men
could be seen and directed through the shower.

231. To reach the position occupied by the theodolite
at the end of our second line, I had to wade breast-deep
through snow which seemed as dry and soft as flour.
The toil of the men upon the glacier in breaking through
the snow was prodigious. But they did not flinch, and
after a time the leader stood behind the farthest stake,
and cried, *Nous avons fini*. I was surprised to hear
him so distinctly, for falling snow had been thought
very deadening to sound. The work was finished, and
I struck my theodolite with a feeling of a general who
had won a small battle.

232. We put the house in order, packed up, and shot
by glissade down the steep slopes of *La Filia* to the
vault of the Arveiron. We found the river feeble, but
not dried up. Many weeks must have elapsed since any
water had been sent down from the surface of the
glacier. But at the setting in of winter the fissures
were in a great measure charged with water; and the
Arveiron of to-day was probably due to the gradual
drainage of the glacier. There was now no danger of
entering the vault, for the ice seemed as firm as marble.
In the cavern we were bathed by blue light. The
strange beauty of the place suggested magic, and put
me in mind of stories about fairy caves which I had read
when a boy. At the source of the Arveiron our winter
visit to the Mer de Glace ends; next morning your
deputy was on his way to London.

§ 33. Winter Motion of the Mer de Glace.

233. Here are the measurements executed in the winter of 1859 :—

LINE No. 1.

Stake	. . 1	2	3	4	5	6	7	8	9	10	11
Inches	. . 7	11	14	13	14	14	16	16	12	12	7

LINE No. II.

Stake	. . 1	2	3	4	5	6	7	8	9	10
Inches	. . 6	10	14	16	16	16	18	17	15	14

234. Thus the winter motion of the Mer de Glace near the Montanvert is, in round numbers, half the summer motion.

235. As in summer, the eastern side of the glacier at this place moved quicker than the western.

§ 34. Motion of the Grindelwald and Aletsch Glaciers.

236. As regards the question of motion, to no other glacier have we devoted ourselves with such thoroughness as to the Mer de Glace; we are, however, able to add a few measurements of other celebrated glaciers. Near the village of Grindelwald in the Bernese Oberland, there are two great ice-streams called respectively the Upper and the Lower Grindelwald glaciers, the second of which is frequently visited by travellers in the Alps. Across it on August 6, 1860, a series of twelve stakes was fixed by Mr. Vaughan Hawkins and myself. Measured on the 8th and reduced to its daily rate, the motion of these stakes was as follows :—

MOTION OF LOWER GRINDELWALD GLACIER.

					East							West
Stake . .	1	2	3	4	5	6	7	8	9	10	11	12
Inches .	18	19	20	21	21	21	22	20	19	18	17	14

237. The theodolite was here planted a little below
the footway leading to the higher glacier region, and at
about a mile above the end of the glacier. The measurement was rendered difficult by crevasses.

238. The largest glacier in Switzerland is the Great
Aletsch, to which further reference shall subsequently
be made. Across it on August 14, 1860, a series of
thirty-four stakes was planted by Mr. Hawkins and me.
Measured on the 16th and reduced to their daily rate,
the velocities were found to be as follows :—

MOTION OF GREAT ALETSCH GLACIER.

			East									
Stake . .	1	2	3	4	5	6	7	8	9	10	11	12
Inches .	2	3	4	6	8	11	13	14	16	17	17	19
Stake . .	13	14	15	16	17	18	19	20	21	22	23	
Inches .	19	18	18	17	19	19	19	19	17	17	15	
Stake . .	24	25	26	27	28	29	30	31	32	33	34	
Inches .	16	17	17	17	17	17	17	17	16	12	12	
												West

239. The maximum motion here is nineteen inches a
day. Probably the eastern side of the glacier is shallow,
the retardation of the bed making the motion of the
eastern stakes inconsiderable. The width of the glacier
here is 9,030 links, or about a mile and a furlong.
The theodolite was planted high among the rocks on
the western flank of the mountain, about half a mile
above the Märgelin See.

§ 35. *Motion of Morteratsch Glacier.*

240. Far to the east of the Oberland and in that
interesting part of Switzerland known as the Ober En-
gadin, stands a noble group of mountains, less in height
than those of the Oberland, but still of commanding
elevation. The group derives its name from its most
dominant peak, the Piz Bernina. To reach the place
we travel by railway from Basel to Zürich, and from
Zürich to Chur (French Coire), whence we pass by diligence over either the Albula pass or the Julier pass to
the village of Pontresina. Here we are in the imme-
diate neighbourhood of the Bernina mountains.

241. From Pontresina we may walk or drive along a
good coach road over the Bernina pass into Italy. At
about an hour above the village you would look from
the road into the heart of the mountains, the line of
vision passing through a valley, in which is couched a
glacier of considerable size. Along its back you would
trace a medial moraine, and you could hardly fail to
notice how the moraine, from a mere narrow streak
at first, widens gradually as it descends, until finally it
quite covers the lower end of the glacier. Nor is this
an effect of perspective; for were you to stand upon the
mountain slopes which nourish the glacier, you would
see thence also the widening of the streak of rubbish,
though the perspective here would tend to narrow
the moraine as it retreats downwards.

2·12. The ice-stream here referred to is the Morteratsch glacier, the end of which is a short, hour's walk from the village of Pontresina. We have now to determine its rate of motion and to account for the widening of its medial moraine. • ·

243. In the summer of 1864 Mr. Hirst and myself set out three lines of stakes across the glacier. The first line crossed the ice high up; the second a good distance lower down, and the third lower still. Even the third line, however, was at a considerable distance above the actual snout of the glacier. The daily motion of these three lines was as follows :—

FIRST LINE.

Stake	. .	1	2	3	4	5	6	7	8	9	10	11
Inches	. .	8	12	13	13	14	13	12	12	10	7	5

SECOND LINE.

Stake	. .	1	2	3	4	5	6	7	8	9	10	11
Inches	. .	1	4	6	8	10	11	11	11	11	11	11

THIRD LINE.

Stake	. .	1	2	3	4	5	6	7	8	9	10	11
Inches	. .	1	2	4	5	6	6	7	7	5	5	4

2·14. Compare these lines together. You notice the velocity of the first is greater than that of the second, and the velocity of the second greater than that of the third.

2·15. The lines were permitted to move downwards for

100 hours, at the end of which time the spaces passed
over by the points of swiftest motion of the three lines
were as follows :—

MAXIMUM MOTION IN 100 HOURS.

First line . . . . 56 inches.
Second line . . . . 45 „
Third line . . . . 30 „

246. Here then is a demonstration that the upper
portions of the Morteratsch glacier are advancing on
the lower ones. *In 1871 the motion of a point on the
middle of the glacier near its snout was found to be less
than two inches a day!*

247. What, then, is the consequence of this swifter
march of the upper glacier?   Obviously to squeeze this
medial moraine longitudinally, and to cause it to spread
out laterally. We have here distinctly revealed the cause
of the widening of the medial moraine.

248. It has been a question much discussed, whether a
glacier is competent to scoop out or deepen the valley
through which it moves, and this very Morteratsch
glacier has been cited to prove that such is not the case.
Observers went to the snout of the glacier, and finding
it sensibly quiescent, they concluded that no scooping
occurred.   But those who contended for the power of
glaciers to excavate valleys never stated, or meant to
state, that it was the snout of the glacier which did
the work.   In the Morteratsch glacier the work of
excavation, which certainly goes on to a greater or

less extent, must be far more effectual high up the valley than at the end of the glacier.

§ 36. *Birth of a Crevasse : Reflections.*

249. Preserving the notion that we are working together, we will now enter upon a new field of enquiry. We have wrapped up our chain, and are turning homewards after a hard day's work upon the Glacier du Géant, when under our feet, as if coming from the body of the glacier, an explosion is heard. Somewhat startled, we look enquiringly over the ice. The sound is repeated, several shots being fired in quick succession. They seem sometimes to our right, sometimes to our left, giving the impression that the glacier is breaking all round us. Still nothing is to be seen.

250. We closely scan the ice, and after an hour's strict search we discover the cause of the reports. They announce the birth of a crevasse. Through a pool upon the glacier we notice air bubbles ascending, and find the bottom of the pool crossed by a narrow crack, from which the bubbles issue. Right and left from this pool we trace the young fissure through long distances. It is sometimes almost too feeble to be seen, and at no place is it wide enough to admit a knife-blade.

251. It is difficult to believe that the formidable fissures among which you and I have so often trodden with awe, could commence in this small way. Such, however, is the case. The great and gaping chasms on

and above the ice-falls of the Géant and the Talèfre begin as narrow cracks, which open gradually to crevasses. We are thus taught in an instructive and impressive way that appearances suggestive of very violent action may really be produced by processes so slow as to require refined observations to detect them. In the production of natural phenomena two things always come into play, the *intensity* of the acting force, and the *time* during which its acts. Make the intensity great, and the time small, and you have sudden convulsion; but precisely the same apparent effect may be produced by making the intensity small, and the time great. This truth is strikingly illustrated by the Alpine ice-falls and crevasses; and many geological phenomena, which at first sight suggest violent convulsion, may be really produced in the selfsame almost imperceptible way.

§ 37. *Icicles.*

252. The crevasses are grandest on the higher névés, where they sometimes appear as long yawning fissures, and sometimes as chasms of irregular outline. A delicate blue light shimmers from them, but this is gradually lost in the darkness of their profounder portions. Over the edges of the chasms, and mostly over the southern edges, hangs a coping of snow, and from this depend like stalactites rows of transparent icicles, 10, 20, 30 feet long. These pendent spears

constitute one of the most beautiful features of the higher crevasses.

253. How are they produced? Evidently by the thawing of the snow. But why, when once thawed, should the water freeze again to solid spears? You have seen icicles pendent from a house-eave, which have been manifestly produced by the thawing of the snow upon the roof. If we understand these, we shall also understand the vaster stalactites of the Alpine crevasses.

254. Gathering up such knowledge as we possess, and reflecting upon it patiently, let us found on it, if we can, a theory of icicles.

255. First, then, you are to know that the *air* of our atmosphere is hardly heated at all by the rays of the sun, whether visible or invisible. The air is highly transparent to all kinds of rays, and it is only the scanty fraction to which it is *not* transparent that expend their force in warming it.

256. Not so, however, with the snow on which the sunbeams fall. It absorbs the solar heat, and on a sunny day you may see the summits of the high Alps glistening with the water of liquefaction. The *air* above and around the mountains may at the same time be many degrees below the freezing point in temperature.

257. You have only to pass from sunshine into shade to prove this. A single step suffices to carry you from a place where the thermometer stands high to one

where it stands low; the change being due, not to any
difference in the temperature of the *air*, but simply to
the withdrawal of the thermometer from the direct
action of the solar rays. Nay, without shifting the
thermometer at all, by interposing a suitable screen,
which cuts off the sun's rays, the coldness of the air
may be demonstrated.

258. Look now to the snow upon your house roof.
The sun plays upon it, and melts it; the water trickles
to the eave and then drops down. If the eave face the
sun the water remains water; but if the eave do not
face the sun, the drop, before its quits its parent snow,
is already in shadow. Now the shaded space, as we
have learnt, may be below the freezing temperature. If
so, the drop, instead of falling, congeals, and the
rudiment of an icicle is formed. Other drops and
driblets succeed, which trickle over the rudiment, congeal upon it in part and *thicken* it at the root. But a
portion of the water reaches the free end of the icicle,
hangs from it, and is there congealed before it escapes.
The icicle is thus *lengthened.* In the Alps, where the
liquefaction is copious and the cold of the shaded
crevasse intense, the icicles, though produced in the
same way, naturally grow to a greater size. The drainage of the snow after the sun's power is withdrawn
also produces icicles.

259. It is interesting and important that you should
be able to explain the formation of an icicle; but it is

fur more important that you should realise the way in which the various threads of what we call Nature are woven together. You cannot fully understand an icicle without first knowing that solar beams powerful enough to fuse the snows and blister the human skin, nay, it might be added, powerful enough, when concentrated, to burn up the human body itself, may pass through the air, and still leave it at an icy temperature.

§ 38. *The Bergschrund.*

260. Having cleared away this difficulty, let us turn once more to the crevasses, taking them in the order of their formation. First then above the névé we have the final Alpine peaks and crests, against which the snow is often reared as a steep buttress. We have already learned that both névés and glaciers are moving slowly downwards; but it usually happens that the attachment of the highest portion of the buttress to the rocks is great enough to enable it to hold on while the lower portion breaks away. A very characteristic crevasse is thus formed, called in the German-speaking portion of the Alps a *Bergschrund*. It often surrounds a peak like a fosse, as if to defend it against the assaults of climbers.

261. Look more closely into its formation. Imagine the snow as yet unbroken. Its higher portions cling to the rocks, and move downwards with extreme slowness. But its lower portions, whether from their

greater depth and weight, or their less perfect attachment, are compelled to move more quickly. *A pull is therefore exerted, tending to separate the lower from the upper snow.* For a time this pull is resisted by the cohesion of the névé; but this at length gives way, and a crack is formed exactly *across* the line in which the pull is exerted. In other words, *a crevasse is formed at right angles to the line of tension.*

§ 39. *Transverse Crevasses.*

262. Both on the névé and on the glacier the origin of the crevasses is the same. Through some cause or other the ice is thrown into a state of strain, and as it cannot *stretch* it *breaks* across the line of tension. Take, for example, the ice-fall of the Géant, or of the Talèfre, above which you know the crevasses yawn terribly. Imagine the névé and the glacier entirely peeled away, so as to expose the surface over which they move. From the Col du Géant we should see this surface falling gently to the place now occupied by the brow of the cascade. Here the surface would fall steeply down to the bed of the present Glacier du Géant, where the slope would become gentle once more.

263. Think of the névé moving over such a surface. It descends from the Col till it reaches the brow just referred to. It crosses the brow, and must bend down to keep upon its bed. Realise clearly what must occur. The surface of the névé is evidently thrown into a

state of strain; it breaks and forms a crevasse. Each fresh portion of the névé as it passes the brow is similarly broken, and thus a succession of crevasses is sent down the fall. Between every two chasms is a great transverse ridge. Through local strains upon the fall those ridges are also frequently broken across, towers of ice—*séracs*—being the result. Down the fall both ridges and séracs are borne, the dislocation being augmented during the descent.

264. What must occur at the foot of the fall? Here the slope suddenly lessens in steepness. It is plain that the crevasses must not only cease to open here, but that they must in whole or in part close up. At the summit of the fall, the bending was such as to make the surface convex; at the bottom of the fall the bending renders the surface concave. In the one case we have *strain*, in the other *pressure*. In the one case, therefore, we have the *opening*, and in the other the *closing* of crevasses. This reasoning corresponds exactly with the facts of observation.

265. Lay bare your arm and stretch it straight. Make two ink dots half an inch or an inch apart, exactly opposite the elbow. Bend your arm, the dots approach each other, and are finally brought together. Let the two dots represent the two sides of a crevasse at the bottom of an ice-fall; the bending of the arm resembles the bending of the ice, and the closing up of the dots resembles the closing of the fissures.

266. The same remarks apply to various portions of the Mer de Glace.   At certain places the inclination changes from a gentler to a steeper slope, and on crossing the brow between both the glacier breaks its back. *Transverse crevasses* are thus formed.   There is such a change of inclination opposite to the Angle, and a still greater but similar change at the head of the Glacier des Bois.   The consequence is that the Mer de Glace at the former point is impassable, and at the latter the rending and dislocation are such as we have seen and described.   Below the Angle, and at the bottom of the Glacier des Bois, the steepness relaxes, the crevasses heal up, and the glacier becomes once more continuous and compact.

§ 40. *Marginal Crevasses.*

267. Supposing, then, that we had no changes of inclination, should we have no crevasses?   We should certainly have less of them, but they would not wholly disappear.   For other circumstances exist to throw the ice into a state of strain, and to determine its fracture. The principal of these is the more rapid movement of the centre of the glacier.

268. Helped by the labours of an eminent man, now dead, the late Mr. Wm. Hopkins, of Cambridge, let us master the explanation of this point together.   But the pleasure of mastering it would be enhanced if we could see beforehand the perplexing and delusive appearances

accounted for by the explanation. Could my wishes be
followed out, I would at this point of our researches carry
you off with me to Basel, thence to Thun, thence to
Interlaken, thence to Grindelwald, where you would
find yourself in the actual presence of the Wetterhorn
and the Eiger, with all the greatest peaks of the Bernese
Oberland, the Finsteraarhorn, the Schreckhorn, the
Monch, the Jungfrau, at hand. At Grindelwald, as we
have already learnt, there are two well-known glaciers
—the Ober Grindelwald and the Unter Grindelwald
glaciers—on the latter of which our observations should
commence.

269. Dropping down from the village to the bottom
of the valley, we should breast the opposite mountain,
and with the great limestone precipices of the Wetterhorn to our left, we should get upon a path which commands a view of the glacier.   Here we should see
beautiful examples of the opening of crevasses at the
summit of a brow, and their closing at the bottom.
But the chief point of interest would be the crevasses
formed at the *side* of this glacier—the *marginal crevasses*,
as they may be called.

270. We should find the side copiously fissured, even
at those places where the centre is compact ; and we
should particularly notice that the fissures would
neither run in the direction of the glacier, nor straight
across it, but that they would be *oblique* to it, enclosing
an angle of about 45 degrees with the sides.   Starting

from the side of the glacier the crevasses would be seen
to point *upwards* ; that is to say, the ends of the fissures
abutting against the bounding mountain would appear
to be *dragged down*. Were you less instructed than
you now are, I might lay a wager that the aspect of
these fissures would cause you to conclude that the
centre of the glacier is left behind by the quicker motion
of the sides.

271. This indeed was the conclusion drawn by M.
Agassiz from this very appearance, before he had
measured the motion of the sides and centre of the
glacier of the Unteraar. Intimately versed with the
treatment of mechanical problems, Mr. Hopkins immediately deduced the obliquity of the lateral crevasses
from the quicker flow of the centre. Standing beside
the glacier with pencil and note-book in hand, I
would at once make the matter clear to you thus.

272. Let A C, in the annexed figure, be one side of
the glacier, and B D the other; and let the direction of

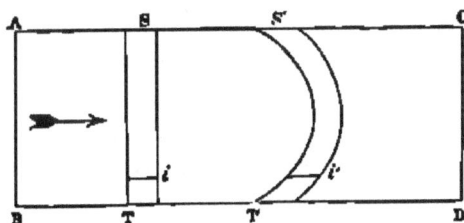

motion be that indicated by the arrow. Let S T be a
transverse slice of the glacier, taken straight across it,

say to-day. A few days or weeks hence this slice will
have been carried down, and because the centre moves
more quickly than the sides it will not remain straight,
but will bend into the form s' τ'.

273. Supposing τ i to be a small square of the
original slice near the side of the glacier. In its new
position the square will be distorted to the lozenge-
shaped figure τ' i'. Fix your attention upon the diagonal
τ i of the square; in the lower position this diagonal,
if the ice could stretch, would be lengthened to τ' i'.
But the ice does not stretch; it breaks, and we have
a crevasse formed at right angles to τ' i'. The mere
inspection of the diagram will assure you that the
crevasse will point obliquely *upwards*.

274. Along the whole side of the glacier the quicker
movement of the centre produces a similar state of
strain; and the consequence is that the sides are
copiously cut by those oblique crevasses, even at places
where the centre is free from them.

275. It is curious to see at other places the transverse
fissures of the centre uniting with those at the sides,
so as to form great curved crevasses which stretch
across the glacier from side to side. The convexity of
the curve is turned *upwards*, as mechanical principles
declare it ought to be. (See sketch on opposite page.)
But if you were ignorant of those principles, you would
never infer from the aspect of these curves the quicker
motion of the centre. In landslips, and in the motion

of partially indurated mud, you may sometimes notice
appearances similar to those exhibited by the ice.

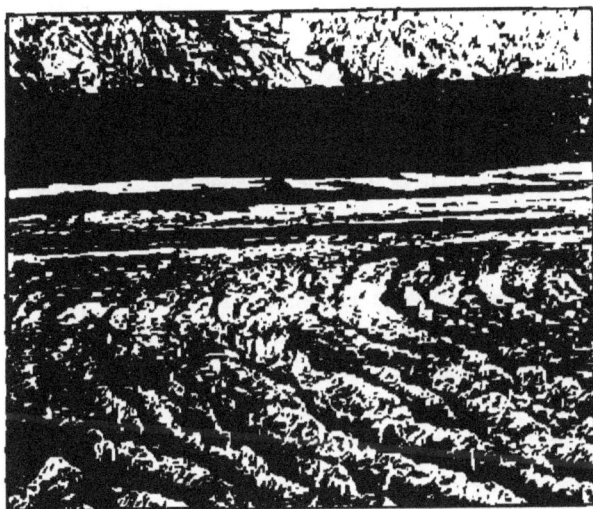

SKETCH OF CURVED CREVASSES; THE GLACIER MOVES FROM LEFT TO RIGHT.

§ 41. *Longitudinal Crevasses.*

276. We have thus unravelled the origin of both
transverse and marginal crevasses. But where a glacier
issues from a steep and narrow defile upon a comparatively level plain which allows it room to expand
laterally, its motion is in part arrested, and the level
portion has to bear the thrust of the steeper portions
behind. Here the line of thrust is in the direction of
the glacier, while the direction at right angles to this

is one of tension. Across this latter the glacier breaks, and *longitudinal crevasses* are formed.

277. Examples of this kind of crevasse are furnished by the lower part of the Glacier of the Rhone, when looked down upon from the Grimsel Pass, or from any commanding point on the flanking mountains.

§ 42. *Crevasses in relation to Curvature of Glacier.*

278. One point in addition remains to be discussed, and your present knowledge will enable you to master it in a moment. You remember at an early period of our researches that we crossed the Mer de Glace from the Chapeau side to the Montanvert side. I then desired you to notice that the Chapeau side of the glacier was more fissured than either the centre or the Montanvert side (75). Why should this be so? Knowing as we now do that the Chapeau side of the glacier moves more quickly than the other; that the point of maximum motion does not lie on the centre but far east of it, we are prepared to answer this question in a perfectly satisfactory manner.

279. Let A B and C D, in the diagram opposite, represent the two curved sides of the Mer de Glace at the Montanvert, and let *m n* be a straight line across the glacier. Let *o* be the point of maximum motion. The mechanical state of the two sides of the glacier may be thus made plain. Supposing the line *m n* to be a straight elastic string with its ends fixed; let it be

grasped firmly at the point o by the finger and thumb, and drawn to o', keeping the distance between o' and

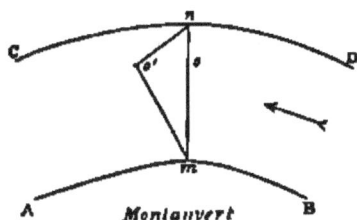

Montanvert

the side c D constant.   Here the length, $n\,o$ of the string would have stretched to $n\,o'$, and the length $m\,o$ to $m\,o'$, and you see plainly that the stretching of the short line, in comparison with its length, is greater than that of the long line in comparison with its length.   In other words, the strain upon $n\,o'$ is greater than that upon $m\,o'$; so that if one of them were to break under the strain, it would be the short one.

280. These two lines represent the conditions of strain upon the two sides of the glacier.   The sides are held back, and the centre tries to move on, a strain being thus set up between the centre and sides.   But the displacement of the point of maximum motion through the curvature of the valley makes the strain upon the eastern ice greater than that upon the western.   The eastern side of the glacier is therefore more crevassed than the western.

281. Here indeed resides the difficulty of getting along the eastern side of the Mer de Glace : a difficulty

which was one reason for our crossing the glacier opposite to the ·Montanvert. There are two convex sweeps on the eastern side to one on the western side, hence on the whole the eastern side of the Mer de Glace is most riven.

§ 43. *Moraine-ridges, Glacier Tables, and Sand Cones.*

282. When you and I first crossed the Mer de Glace from Trélaporte to the Couvercle, we found that the stripes of rocks and rubbish which constituted the medial moraines were ridges raised above the general level of the glacier to a height at.some places of twenty or thirty feet. On examining these ridges we found the rubbish to be superficial, and that it rested upon a great spine of ice which ran along the back of the glacier. By what means has this ridge of ice been raised?

283. Most boys have read the story of Dr. Franklin's placing bits of cloth of various colours upon snow on a sunny day. The bits of cloth sank in the snow, the dark ones most.

284. Consider this experiment. The sun's rays first of all fall upon the upper surface of the cloth and warm it. The heat is then conducted through the cloth to the under surface, and the under surface passes it on to the snow, which is finally liquefied by the heat. It is quite manifest that the quantity of snow melted will altogether depend upon the amount of heat sent from the upper to the under surface of the cloth.

285. Now cloth is what is called a bad conductor. It does not permit heat to travel freely through it. But where it has merely to pass through the thickness of a single bit of cloth, a good quantity of the heat gets through. But if you double or treble or quintuple the thickness of the cloth; or, what is easier, if you put several pieces one upon the other, you come at length to a point where no sensible amount of heat could get through from the upper to the under surface.

286. What must occur if such a thick piece, or such a series of pieces of cloth, were placed upon snow on which a strong sun is falling? The snow round the cloth is melted, but that underneath the cloth is protected. If the action continue long enough the inevitable result will be, that the level of the snow all round the cloth will sink, and the cloth will be left behind perched upon an eminence of snow.

287. If you understand this, you have already mastered the cause of the moraine-ridges. They are not produced by any swelling of the ice upwards. But the ice underneath the rocks and rubbish being protected from the sun, the glacier right and left melts away and leaves a ridge behind.

288. Various other appearances upon the glacier are accounted for in the same way. Here upon the Mer de Glace we have flat slabs of rock sometimes lifted up on pillars of ice. These are the so-called *Glacier Tables*. They are produced, not by the growth of a stalk of ice

I

out of the glacier, but by the melting of the glacier all
round the ice protected by the stone.   Here is a sketch
of one of the Tables of the Mer de Glace.

289. Notice moreover that a glacier table is hardly ever
set square upon its pillar.   It generally leans to one
side, and repeated observation teaches you that it so
leans as to enable you always to draw the north and
south line upon the glacier.   For the sun being south
of the zenith at noon pours its rays against the southern end of the table, while the northern end remains in
shadow.   The southern end, therefore, being most
warmed does not protect the ice underneath it so effectually as the northern end.   The table becomes inclined,
and ends by sliding bodily off its pedestal.

290. In the figure opposite we have what may be called

an ideal Table. The oblique lines represent the direction of the sunbeams, and the consequent tilting of the table here shown resembles that observed upon the glaciers.

291. A pebble will not rise thus: like Franklin's single bit of cloth, a dark-coloured pebble sinks in the ice. A spot of black mould will not rest upon the surface, but will sink; and various parts of the Glacier du Géant are honeycombed by the sinking of such spots of dirt into the ice.

292. But when the dirt is of a thickness sufficient to protect the ice the case is different.   Sand is often

washed away by a stream from the mountains, or from the moraines, and strewn over certain spaces of the glacier.   A most curious action follows: the sanded

surface rises, the part on which the sand lies thickest
rising highest. Little peaks and eminences jut forth,
and when the distribution of the sand is favourable,
and the action sufficiently prolonged, you have little
mountains formed, sometimes singly, and sometimes
grouped so as to mimic the Alps themselves. The *Sand
Cones* of the Mer de Glace are not striking; but on
the Görner, the Aletsch, the Morteratsch, and other
glaciers, they form singly and in groups, reaching
sometimes a height of ten or twenty feet.

§ 44. *The Glacier Mills or Moulins.*

293. You and I have learned by long experience the
character of the Mer de Glace. We have marched over
it daily, with a definite object in view, but we have
not closed our eyes to other objects. It is from side
glimpses of things which are not at the moment occu-
pying our attention that fresh subjects of enquiry arise
in scientific investigation.

294. Thus in marching over the ice near Trélaporte
we were often struck by a sound resembling low rumbling
thunder. We subsequently sought out the origin of
this sound, and found it.

295. A large area of this portion of the glacier is
unbroken. Driblets of water have room to form rills;
rills to unite and form streams ; streams to combine to
form rushing brooks, which sometimes cut deep chan
nels in the ice. Sooner or later these streams reach a

strained portion of the glacier, where a crack is formed
across the stream.   A way is thus opened for the water
to the bottom of the glacier.   By long action the stream
hollows out a shaft, the crack thus becoming the
starting-point of a funnel of unseen depth, into which
the water leaps with the sound of thunder.

296. This funnel and its cataract form a glacier Mill
or *Moulin.*

297. Let me grasp your hand firmly while you stand
upon the edge of this shaft and look into it.   The hole,
with its pure blue shimmer, is beautiful, but it is terrible.
Incautious persons have fallen into these shafts, a
second or two of bewilderment being followed by sudden
death.   But caution upon the glaciers and mountains
ought, by habit, to be made a second nature to explorers
like you and me.

298. The crack into which the stream first descended
to form the moulin, moves down with the glacier.   A
succeeding portion of the ice reaches the place where
the breaking strain is exerted.   A new crack is then
formed above the moulin, which is thenceforth forsaken by the stream, and moves downward as an empty
shaft.   Here upon the Mer de Glace, in advance of the
Grand Moulin, we see no less than six of these forsaken
holes.   Some of them we sound to a depth of 90 feet.

299. But you and I both wish to determine, if possible,
the entire depth of the Mer de Glace.   The Grand
Moulin offers a chance of doing this which we must not

neglect. Our first effort to sound the moulin fails through the breaking of our cord by the impetuous plunge of the water. A lump of grease in the hollow of a weight enables a mariner to judge of a sea bottom. We employ such a weight, but cannot reach the bed of the glacier. A depth of 163 feet is the utmost reached by our plummet.

300. From July 28 to August 8 we have watched the progress of the Grand Moulin. On the former date the position of the Moulin was fixed. On the 31st it had moved down 50 inches; a little more than a day afterwards it had moved 74 inches. On August 8 it had moved 198 inches, which gives an average of about 18 inches in twenty-four hours. No doubt next summer upon the Mer de Glace a Grand Moulin will be found thundering near Trélaporte; but like the crevasse of the Grand Plateau, already referred to (§ 16), it will not be our Moulin. This, or rather the ice which it penetrated, is now probably more than a mile lower down than it was in 1857.

§ 45. *The Changes of Volume of Water by Heat and Cold.*

301. We have noticed upon the glacier shafts and pits filled with water of the most delicate blue. In some cases these have been the shafts of extinct moulins closed at the bottom. A theory has been advanced to account for them, which, though it may be untenable, opens out considerations regarding the

properties of water that ought to be familiar to enquirers like you and me.

302. In our dissection of lake ice by a beam of heat (§ 11) we noticed little vacuous spots at the centres of the liquid flowers formed by the beam. These spots we referred to the fact that when ice is melted the water produced is less in volume than the ice, and that hence the water of the flower was not able to occupy the whole space covered by the flower.

303. Let us more fully illustrate this subject. Stop a small flask water-tight with a cork, and through the cork introduce a narrow glass tube also water-tight. It is easy to fill the flask with water so that the liquid shall stand at a certain height in the glass tube.

304. Let us now warm the flask with the flame of a spirit-lamp. On first applying the flame you notice a momentary sinking of the liquid in the glass tube. This is due to the momentary expansion of the flask by heat; it becomes suddenly larger when the flame is first applied.

305. But the expansion of the water soon overtakes that of the flask and surpasses it. We immediately see the rise of the liquid column in the glass tube, exactly as mercury rises in the tube of a warmed thermometer.

306. Our glass tube is ten inches long, and at starting the water stood in it at a height of five inches. We will apply the spirit-lamp flame until the water rises quite to the top of the tube and trickles over. This

experiment suffices to show the expansion of the water by heat.

307. We now take a common finger-glass and put into it a little pounded ice and salt. On this we place the flask, and then build round it the freezing mixture. The liquid column retreats down the tube, proving the contraction of the liquid by cold. We allow the shrinking to continue for some minutes, noticing that the downward retreat of the liquid becomes gradually slower, and that it finally ceases altogether.

308. Keep your eye upon the liquid column; it remains quiescent for a fraction of a minute, and then moves once more. But its motion is now *upwards* instead of downwards. *The freezing mixture now acts exactly like the flame.*

309. It would not be difficult to pass a thermometer through the cork into the flask, and it would tell us the exact temperature at which the liquid ceased to contract and began to expand. At that moment we should find the temperature of the liquid a shade over 39° Fahr.

310. At this temperature, then, water attains *its maximum density.*

311. Seven degrees below this temperature, or at 32° Fahr., the liquid begins to turn into solid crystals of ice, which you know swims upon water because it is bulkier for a given weight. In fact, this halt of the approaching molecules at the temperature of 39°, is but the preparation for the subsequent act of crystallisation, in which

the expansion by cold culminates. Up to the point of
solidification the increase of volume is slow and gradual;
while in the act of solidification it is sudden, and of
overwhelming strength.

312. By this force of expansion the Florentine Acade-
micians long ago burst a sphere of copper nearly three
quarters of an inch in thickness. By the same force
the celebrated astronomer Huyghens burst in 1667 iron
cannons a finger breadth thick. Such experiments
have been frequently made since. Major Williams
during a severe Quebec winter filled a mortar with
water, and closed it by driving into its muzzle a plug
of wood. Exposed to a temperature 50° Fahr. below
the freezing point of water, the metal resisted the
strain, but the plug gave way, being projected to a
distance of 400 feet. At Warsaw howitzer shells
have been thus exploded; and you and I have shi-
vered thick bomb-shells to fragments, by placing them
for half an hour in a freezing mixture.

313. The theory of the shafts and pits referred to
at the beginning of this section is this:—The water
at the surface of the shaft is warmed by the sun, say to
a temperature of 39° Fahr. The water at the bottom,
in contact with the ice, must be at 32° or near it. The
heavier water is therefore at the top; it will descend
to the bottom, melt the ice there, and thus deepen the
shaft.

314. The circulation here referred to undoubtedly

goes on, and some curious effects are due to it; but
not, I think, the one here ascribed to it. The *deepening*
of a shaft implies a quicker melting of its bottom than
of the surface of the glacier. It is not easy to see how
the fact of the solar heat being first absorbed by water,
and then conveyed by it to the bottom of the shaft,
should make the melting of the bottom more rapid than
that of the ice which receives the direct impact of
thé solar rays. The surface of the glacier must sink
at least as rapidly as the bottom of the pit, so that
the circulation, though actually existing, cannot produce
the effect ascribed to it.

§ 46. *Consequences flowing from the foregoing Properties
of Water. Correction of Errors.*

315. I was not much above your age when the
property of water ceasing to contract by cold at a
temperature of 39° Fahr. was made known to me, and I
still remember the impression it made upon me. For
I was asked to consider what would occur in case this
solitary exception to an otherwise universal law ceased
to exist.

316. I was asked to reflect upon the condition of a
lake stored with fish and offering its surface to very
cold air. It was made clear to me that the water on
being first chilled would shrink in volume and become
heavier, that it would therefore sink and have its place

supplied by the warmer and lighter water from the
deeper portions of the lake.

317. It was pointed out to me that without the law
referred to this process of circulation would go on until
the whole water of the lake had been lowered to the
freezing temperature. Congelation would then begin,
and would continue as long as any water remained to be
solidified. One consequence of this would be to destroy
every living thing contained in the lake. Other calamities
were added, all of which were said to be prevented by
the perfectly exceptional arrangement, that after a certain time the *colder* water becomes the *lighter*, floats
on the surface of the lake, is there congealed, thus
throwing a protecting roof over the life below. .

318. Count Rumford, one of the most solid of scientific
men, writes in the following strain about this question :
—' It does not appear to me that there is anything
which human sagacity can fathom, within the wide-
extended bounds of the visible creation, which affords a
more striking or more palpable proof of the wisdom of
the Creator, and of the special care He has taken in the
general arrangement of the universe, to preserve animal
life, than this wonderful contrivance.

319. ' Let me beg the attention of my readers while I
endeavour to investigate this most interesting subject ;
and let me at the same time bespeak his candour and
indulgence. I feel the danger to which a mortal exposes himself who has the temerity to explain the designs

of Infinite Wisdom. The enterprise is adventurous,
but it surely cannot be improper.

320. 'Had not Providence interfered on this occasion
in a manner which may well be considered as *miracu-
lous*, all the fresh water within the polar circle must
inevitably have been frozen to a very great depth in
winter, and every plant and tree destroyed.'

321. Through many pages of his book Count Rumford
continues in this strain to expound the ways and in-
tentions of the Almighty, and he does not hesitate to
apply very harsh words to those who cannot share his
notions. He calls them hardened and degraded. We
are here warned of the fact, which is too often forgotten, that the pleasure or comfort of a belief, or the
warmth or exaltation of feeling which it produces, is no
guarantee of its truth. For the whole of Count Rumford's delight and enthusiasm in connexion with this
subject, and the whole of his ire against those who did
not share his opinions, were founded upon an erroneous
notion.

322. Water is *not* a solitary exception to an otherwise
general law. There are other molecules than those of
this liquid which require more room in the solid crystalline condition than in the adjacent molten condition.
Iron is a case in point. Solid iron floats upon molten
iron exactly as ice floats upon water. Bismuth is a still
more impressive case, and we could shiver a bomb as
certainly by the solidification of bismuth as by that

of water.   There is no fish to be taken care of here,
still the 'contrivance' is the same.

323. I am reluctant to mention them in the same
breath with Count Rumford, but I am told that in our
own day there are people who profess to find the comforts
of a religion in a superstition lower than any that has
hitherto degraded the civilized human mind. So that
the *happiness* of a faith and the *truth* of a faith are two
totally different things.

.324. 'Life and the conditions of life are in necessary
harmony. This is a truism, for without the suitable
conditions life could not exist. But both life and its
conditions set forth the operations of inscrutable Power.
We know not its origin; we know not its end. And
the presumption, if not the degradation, rests with
those who place upon the throne of the universe a
magnified image of themselves, and make its doings
a mere colossal imitation of their own.

§ 47. *The Molecular Mechanism of Water-congelation.*

325. But let us return to our science. How are we
to picture this act of expansion on the part of freezing
water?   By what operation do the molecules demand
with such irresistible emphasis more room in the solid
than in the adjacent liquid condition? In all cases of
this kind we must derive our conceptions from the
world of the senses, and transfer them afterwards to a
world transcending the range of the senses.

326. You have not forgotten our conversation regarding 'atomic poles' (§ 10), and how the notion of polar force came to be applied to crystals. With this fresh in your memory, you will have no great difficulty in understanding how expansion of volume may accompany the act of crystallisation.

327. I place a number of magnets before you. They, as matter, are affected by gravity, and, if perfectly free, they would move towards each other in obedience to the attraction of gravity.

328. But they are not only matter, but *magnetic* matter. They not only act upon each other by the simple force of gravity, but by the polar force of magnetism. Imagine them placed at a distance from each other, and perfectly free to move. Gravity first makes itself felt and draws them together. For a time the magnetic force issuing from the poles is insensible ; but when a certain nearness is attained, the polar force comes into play. The mutually attracting points close up, the mutually repellent points retreat, and it is easy to see that this action may produce an arrangement of the magnets which requires more room. Suppose them surrounded by a box which exactly encloses them at the moment the polar force first comes into play. It is easy to see that in arranging themselves subsequently the repelled corners and ends of the magnets may be caused to press against the sides of the box, and even to burst it, if the forces be sufficiently strong.

329. Here then we have a conception which may be applied to the molecules of water. They, like the magnets, are acted upon by two distinct forces. For a time while the liquid is being cooled they approach each other, in obedience to their general attraction for each other. But at a certain point new forces, some attractive, some repulsive, *emanating from special points* of the molecules, come into play. The attracted points close up, the repelled points retreat. Thus the molecules turn and rearrange themselves, demanding, as they do so, more space, and overcoming all ordinary resistance by the energy of their demand. This, in general terms, is an explanation of the expansion of water in solidifying: it would be easy to construct an apparatus for its illustration.

, § 48. *The Dirt Bands of the Mer de Glace.*

330. Pass from bright sunshine into a moderately lighted room; for a time all appears so dark that the objects in the room are not to be clearly distinguished. Hit violently by the waves of light (§ 3) the optic nerve is numbed, and requires time to recover its sensitiveness.

331. It is for this reason that I choose the present hour for a special observation on the Mer de Glace. The sun has sunk behind the ridge of Charmoz, and the surface of the glacier is in sober shade. The main portion of our day's work is finished, but we have still sufficient energy to climb the slopes adjacent to the

Montanvert to a height of a thousand feet or thereabouts above the ice.

332. We now look fairly down upon the glacier, and

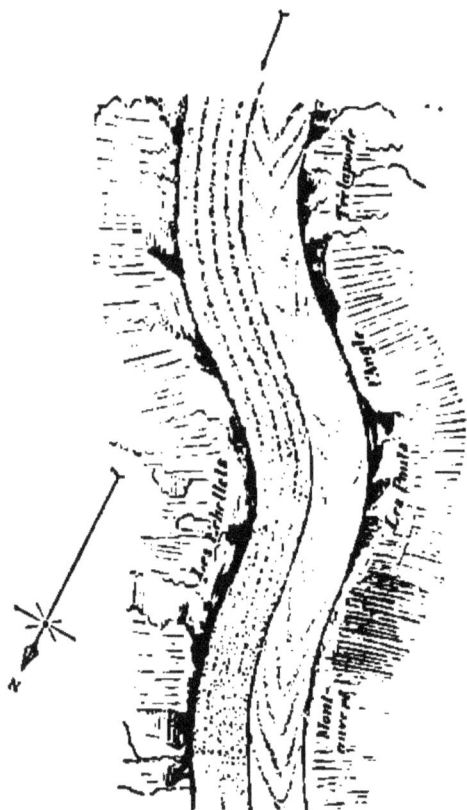

see it less foreshortened than from the Montanvert. We notice the dirt overspreading its eastern side, due to

the crowding together of its medial moraines. We see
the comparatively clean surface of the Glacier du
Géant; but we notice upon this surface an appearance which we have not hitherto seen. It is

crossed by a series of grey bent bands, which follow
each other in succession, from Trélaporte downwards.
We count eighteen of these from our present position.
(See sketch, page 128.)

K

3:):). These are the *Dirt Bands* of the Mer de Glace;
they were first observed by Professor Forbes in 1842.

3:)4. They extend down the glacier further than we
can see; and if we cross the valley of Chamonni, and

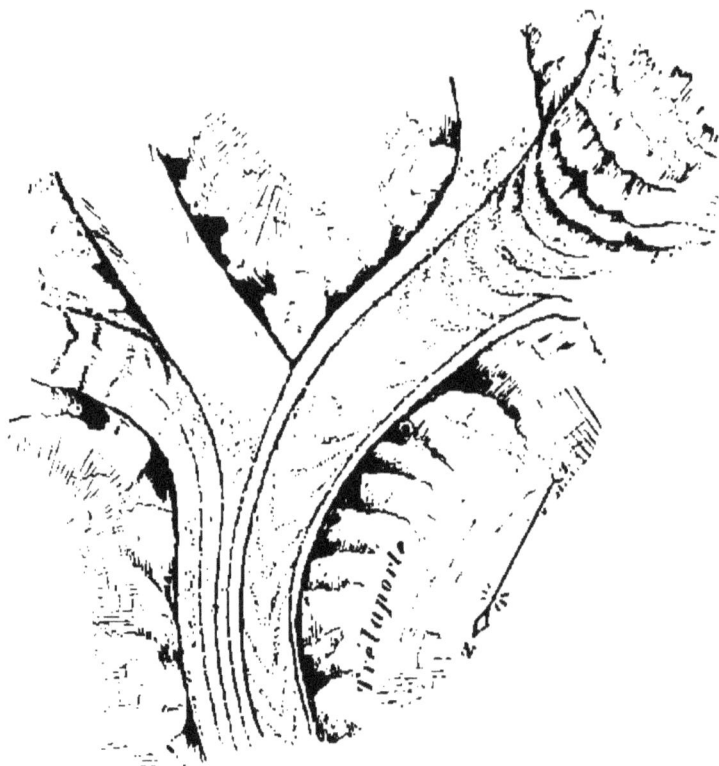

climb the mountains at the opposite side, to a point
near the little unberge, called La Flégère, we shall

command a view of the end of the glacier and observe the completion of the series of bands.   We notice that they are confined throughout to the portion of the glacier derived from the Col du Géant. (See sketch, page 129.) 335. We must trace them to their source.   You know how noble and complete a view is obtained of the glacier and Col du Géant from the Cleft Station above Trélaporte.   Thither we must once more climb; and thence we can see the succession of bands stretching downwards to the Montanvert, and upwards to the base of the ice-cascade upon the Glacier du Géant.   The cascade is evidently concerned in their formation.   (See sketch opposite.)

336. And how?   Simply enough.   The glacier, as we know, is broken transversely at the summit of the ice-fall, and descends the declivity in a series of great transverse ridges.   At the base of the fall, the chasms are closed, but the ridges in part remain forming protuberances, which run like vast wrinkles across the glacier.   These protuberances are more and more bent because of the quicker motion of the centre, and the depressions between them form receptacles for the fine mud and débris washed by the little rills from the adjacent slopes.

337. The protuberances sink gradually through the wasting action of the sun, so that long before Trélaporte is reached they have wholly disappeared.   Not so the dirt of which they were the collectors: it continues to

occupy, in transverse bands, the flat surface of the glacier. At Trélaporte, moreover, where the valley becomes narrow, the bands are much sharpened, obtaining there the character which they afterwards preserve throughout the Mer de Glace. Other glaciers with cascades also exhibit similar bands.

§ 49. *Sea Ice and Icebergs.*

338. We are now equipped intellectually for a campaign into another territory. Water becomes heavier and more difficult to freeze when salt is dissolved in it. Sea water is therefore heavier than fresh, and the Greenland Ocean requires to freeze it a temperature 3½ degrees lower than fresh water. When concentrated till its specific gravity reaches 1·1045, sea water requires for its congelation a temperature 18½ degrees lower than the ordinary freezing-point.*

339. But even when the water is saturated with salt, the crystallising force studiously rejects the salt, and devotes itself to the congelation of the water alone. Hence the ice of sea water, when melted, produces fresh water. The only saline particles existing in such ice are those entangled mechanically in its pores. They have no part or lot in the structure of the crystal.

340. This *exclusiveness*, if I may use the term, of the water molecules; this entire rejection of all foreign

* Scoresby.

elements from the edifices which they build, is enforced to a surprising degree. Sulphuric acid has so strong an affinity for water that it is one of the most powerful agents known to the chemist for the removal of humidity from air. Still, as shown by Faraday, when a mixture of sulphuric acid and water is frozen, the crystal formed is perfectly sweet and free from acidity. The water alone has lent itself to the crystallising force.

341. Every winter in the Arctic regions the sea freezes, roofing itself with ice of enormous thickness and vast extent. By the summer heat, and the tossing of the waves, this is broken up; the fragments are drifted by winds and borne by currents. They clash, they crush each other, they pile themselves into heaps, thus constituting the chief danger encountered by mariners in the polar seas.

342. But among the drifting masses of flat sea-ice, vaster masses sail, which spring from a totally different source. These are the *Icebergs* of the Arctic seas. They rise sometimes to an elevation of hundreds of feet above the water, while the weight of ice submerged is about seven times that seen above.

343. The first observers of striking natural phenomena generally allow wonder and imagination more than their due place. But to exclude all error arising from this cause, I will refer to the journal of a cool and intrepid Arctic navigator, Sir Leopold McClintock. He describes an iceberg 250 feet high, which was aground

in 500 feet of water. This would make the entire height of the berg 750 feet, not an unusual altitude for the greater icebergs.

344. From Baffin's Bay these mighty masses come sailing down through Davis' Straits into the broad Atlantic. A vast amount of heat is demanded for the simple liquefaction of ice (§ 48); and the melting of icebergs is on this account so slow, that when large they sometimes maintain themselves till they have been drifted 2000 miles from their place of birth.

345. What is their origin? The Arctic glaciers. From the mountains in the interior the indurated snows slide into the valleys and fill them with ice. The glaciers thus formed move like the Swiss ones, incessantly downward. But the Arctic glaciers reach the sea, enter it, often ploughing up its bottom into submarine moraines. Undermined by the lapping of the waves, and unable to resist the strain imposed by their own weight, they break across, and discharge vast masses into the ocean. Some of these run aground on the adjacent shores, and often maintain themselves for years. Others escape southward, to be finally dissolved in the warm waters of the Atlantic. The first engraving on the opposite page is copied from a photograph taken by Mr. Bradford during a recent expedition to the Northern seas. The second represents a mass of ice upon the Glacier des Bossons. Their likeness suggests their common origin.

§ 50. *The Æggischhorn, the Märgelin See and its Icebergs.*

346. I am, however, unwilling that you should quit Switzerland without seeing such icebergs as it can show, and indeed there are other still nobler glaciers than the Mer de Glace with which you ought to be acquainted. In tracing the Rhone to its source, you have already ascended the valley of the Rhone. Let us visit it again together; halt at the little town of Viesch, and go from it straight up to the excellent hostelry on the slope of the Æggischhorn. This we shall make our head-quarters while we explore that monarch of European ice-streams, —the great Aletsch glacier.

347. Including the longest of its branches, this noble ice-river is about twenty miles long, while at the middle of its trunk it measures nearly a mile and a quarter from side to side. The grandest mountains of the Bernese Oberland, the Jungfrau, the Monch, the Trugberg, the Aletschhorn, the Breithorn, the Gletscherhorn, and many another noble peak and ridge, are the collectors of its névés. From three great valleys formed in the heart of the mountains these névés are poured, uniting together to form the trunk of the Aletsch at a place named by a witty mountaineer, the 'Place de la Concorde of Nature.' If the phrase be meant to convey the ideas of tranquil grandeur, beauty of form, and purity of hue, it is well bestowed.

348. Our hotel is not upon the peak of the Æggisch-horn, but a brisk morning walk soon places us upon the top. Thence we see the glacier like a broad river stretching upwards to the roots of the Jungfrau, and downwards past the Bel Alp towards its end. Pro-longing the vision downwards, we strike the noblest mountain group in all the Alps,—the Dom and its attendant peaks, the Matterhorn and the Weisshorn. The scene indeed is one of impressive grandeur, a mul-titude of peaks and crests here unnamed contributing to its glory.

349. But low down to our right, and surrounded by the sheltering mountains, is an object the beauty of which startles those who are unprepared for it. Yonder we see the naked side of the glacier, exposing glistening ice-cliffs sixty or seventy feet high. It would seem as if the Aletsch here were engaged in the vain attempt to thrust an arm through a lateral valley. It once did so; but the arm is now incessantly broken off close to the body of the glacier, a great space formerly covered by the ice being occupied by its water of liquefaction. A lake of the loveliest blue is thus formed, which reaches quite to the base of the ice-cliffs, saps them, as the Arctic waves sap the Greenland glaciers, and receives from them the broken masses which it has undermined. As we look down upon the lake, small icebergs sail over the tranquil surface, each resembling a snowy swan accompanied by its shadow.

350. This is the beautiful little lake of Märgelin, or, as the Swiss here call it, the Märgelin See. You see that splash, and immediately afterwards hear the sound of the plunging ice. The glacier has broken before our eyes, and dropped an iceberg into the lake. All over the lake the water is set in commotion, thus illustrating on a small scale the swamping waves produced by the descent of vast islands of ice from the Arctic glaciers. Look to the end of the lake. It is cumbered with the remnants of icebergs now aground, which have been in part wafted thither by the wind, but in part slowly borne by the water which moves gently in this direction.

351. Imagine us below upon the margin of the lake, as I happened to be on one occasion. There is one large and lonely iceberg about the middle. Suddenly a sound like that of a cataract is heard; we look towards the iceberg and see water teeming from its sides. Whence comes the water? the berg has become top-heavy through the melting underneath; it is in the act of performing a somersault, and in rolling over carries with it a vast quantity of water, which rushes like a waterfall down its sides. And notice that the iceberg, which a moment ago was snowy-white, now exhibits the delicate blue colour characteristic of compact ice. It will soon, however, be rendered white again by the action of the sun. The vaster icebergs of the Northern seas sometimes roll over in the same fashion. A week may be spent with delight and profit at the Æggischhorn.

§ 51. *The Bel Alp.*

352. From the Æggischhorn I might lead you along the mountain ridge by the Betten See, the fish of which we have already tasted, to the Rieder Alp, and thence across the Aletsch to the Bel Alp. This is a fine mountain ramble, but you and I prefer making the glacier our highway downwards. Easy at some places, it is by no means child's play at others to unravel its crevasses. But the steady constancy and close observation which we have hitherto found availing in difficult places do not forsake us here. We clear the fissures; and, after four hours of exhilarating work, we find ourselves upon the slope leading up to the Bel Alp hotel.

353. This is one of the finest halting-places in the Alps. Stretching before us up to the Æggischhorn and Märgelin See is the long last reach of the Aletsch, with its great medial moraine running along its back. At hand is the wild gorge of the Massa, in which the snout of the glacier lies couched like the head of a serpent. The beautiful system of the Oberaletsch glaciers is within easy reach. Above us is a peak called the Sparrenhorn, accessible to the most moderate climber, and on the summit of which little more than an hour's exertion will place you and me. Below us now is the Oberaletsch glacier, exhibiting the most perfect of medial moraines. Near us is the great mass of the Aletschhorn, clasped by its névés, and culminating in brown

rock. It is supported by other peaks almost as noble
as itself. The Nesthorn is at hand; while sweeping
round to the west we strike the glorious triad already
referred to, the Weisshorn, the Matterhorn, and the
Dom. Take one glance at the crevasses of the glacier
immediately below us. It tumbles at its end down a
steep incline, and is greatly riven. But the crevasses
open before the steep part is reached, and you notice
the coalescence of marginal and transverse crevasses,
producing a system of curved fissures with the convex-
ities of the curves pointing upwards. The mechanical
reason of this is now known to you. The glacier-tables
are also numerous and fine. I should like to linger
with you here for a week, exploring the existing glaciers,
and tracing out the evidences of others that have passed
away.

§ 52. *The Riffelberg and Görner Glacier.*

854. And though our measurements and observations
on the Mer de Glace are more or less representative of
all that can be made or solved elsewhere, I am unwilling
to leave you unacquainted with the great system of
glaciers which stream from the northern slopes of
Monte Rosa and the adjacent mountains. From the
Bel Alp we can descend to Brieg, and thence drive to
Visp; but you and I prefer the breezy heights, so we
sweep round the promontory of the Nessel, until we
stand over the Rhone valley, in front of Visp. From

this village an hour's walking carries us to Stalden, where the valley divides into two branches: the one leading through Saas over the Monte Moro, and the other through St. Nicholas to Zermatt. The latter is our route.

355. We reach Zermatt, but do not halt there. On the mountain ridge, 4,000 feet above the valley, we discern the Riffelberg hotel. This we reach. Right in front of us is the pinnacle of the Matterhorn, upon the top of which it must appear incredible to you that a human foot could ever tread. Constancy and skill, however, accomplished this, but in the first instance at a terrible price. In the little churchyard of Zermatt we have seen the graves of two of the greatest mountaineers that Savoy and England have produced; and who, with two gallant young companions, fell from the Matterhorn in 1865.

356. At the Riffelberg we are within an hour's walk of the famous Görner Grat, which commands so grand a view of the glaciers of Monte Rosa. But yonder huge knob of perfectly bare rock, which is called the Riffelhorn, must be our station. What the Cleft Station is to the Mer de Glace, the Riffelhorn is to the Görner glacier and its tributaries. From its lower side the rock, easy as it may seem, is inaccessible. Here, indeed, in 1865, a fifth good man met his end, and he also lies beside his fellow countrymen in the churchyard of Zermatt. Passing a little tarn, or lake, called the Riffel See,

we assail the Riffelhorn on its upper side. It is capital
rock-practice to reach the summit; and from it we
command a most extraordinary scene.

357. The huge and many-peaked mass of Monte
Rosa faces us, and we scan its snows from bottom to
top. To the right is the mighty ridge of the Lyskamm,
also laden with snow; and between both lies the
Western Glacier of Monte Rosa. This glacier meets
another from the vast snow-fields of the Cima di Jazzi;
they join to form the Görner glacier, and from their place
of junction stretches the customary medial moraine.
On this side of the Lyskamm rise two beautiful snowy
eminences, the Twins Castor and Pollux; then come
the brown crags of the Breithorn, then the Little Matterhorn, and then the broad snow-field of the Théodule,
out of which springs the Great Matterhorn, and which
you and I will cross subsequently into Italy.

358. The valleys and depressions between these mountains are filled with glaciers. Down the flanks of the
Twin Castor comes the Glacier des Jumeaux, from Pollux
comes the Schwartze glacier, from the Breithorn the
Trifti glacier, then come the Little Matterhorn glacier
and the Théodule glacier, each, as it welds itself to the
trunk, carrying with it its medial moraine. We can
count nine such moraines from our present position.
And to a still more surprising degree than on the Mer
de Glace, we notice the power of the ice to yield to
pressure; the broad névés being squeezed on the trunk

of the Görner into white stripes, which become ever
narrower between their bounding moraines, and finally
disappear under their own shingle.

359. On the two main tributaries we also notice
moraines which seem in each case to rise from the
body of the glacier, appearing in the middle of the

THE GORNER GLACIER, WITH MONTE ROSA IN THE DISTANCE, AND THE
RIFFELHORN TO THE LEFT.

ice without any apparent origin higher up.   These
at their sources, are sub-glacial moraines, which have
been rubbed away from rocky promontories entirely
covered with ice. They lie hidden for a time in the
body of the glacier, and appear at the surface where
the ice above them has been melted away by the sun.

360. This is the place to mention a notion long entertained by the inhabitants of the high Alps, that glaciers possess the power of thrusting out all impurities from them. On the Mer de Glace you and I have noticed large patches of clay and black mud which evidently came from the body of the glacier, and we can therefore understand how natural was this notion of extrusion to people unaccustomed to close observation. But the power of the glacier in this respect is in reality the power of the sun, which fuses the ice above concealed impurities, and, like the bodies of the guides on the Glacier des Bossons (143), brings them to the light of day.

361. On no other glacier will you find more objects of interest than on the Görner. Sand cones, glacier-tables, deep ice gorges cut by streams and bridged fantastically by boulders, moulins, sometimes arched ice-caverns of extraordinary size and beauty. On the lower part of the glacier we notice the partial disappearance of the medial moraine in the crevasses, and its reappearance at the foot of the incline. For many years this glacier was steadily advancing on the meadow in front of it, ploughing up the soil and overturning the châlets in its way. It now shares in the general retreat exhibited during the last fifteen years among the glaciers of the Alps. As usual, a river, the Visp, rushes from a vault at the extremity of the Görner glacier.

§ 53. *Ancient Glaciers of Switzerland.*

362. You have not lost the memory of the old
Moraine, which interested us so much in our first ascent
from the source of the Arveiron; for it opened our
minds to the fact that at one period of its history the
Mer de Glace attained far greater dimensions than it
now exhibits. Our experience since that time has
enabled us to pursue these evidences of ice action to an
extent of which we had then no notion.

363. Close to the existing glacier, for example, we
have repeatedly seen the mountain side laid bare by the
retreat of the ice. This is especially conspicuous just
now, because for the last fifteen or sixteen years the
glaciers of the Alps have been steadily shrinking; so
that it is no uncommon thing to see the marginal rocks
laid bare for a height of fifty, sixty, eighty, or even one
hundred feet above the present glacier. On the rocks
thus exposed we see the evident marks of the sliding;
and our eyes and minds have been so educated in the
observation of these appearances that we are now able
to detect, with certainty, icemarks, or moraines, ancient
or modern, wherever they appear.

364. But the elevations at which we have found such
evidence might well shake belief in the conclusions to
which they point. Beside the Massa Gorge, at 1,000
feet above the present Aletsch, we found a great old
moraine. Descending the meadows between the Bel Alp

L

and Platten, we found another, now clothed with grass, and bearing a village on its back. But I wish to carry you to a region which exhibits these evidences on a still grander and more impressive scale. We have already taken a brief flight to the valley of Hasli and the Glacier of the Aar. Let us make that glacier our starting-point. Walking from it downwards towards the Grimsel, we pass everywhere over rocks singularly rounded, and fluted, and scarred. These appearances are manifestly the work of the glacier in recent times. But we approach the Grimsel, and at the turning of the valley stand before the precipitous granite flank of the mountain. The traces of the ancient ice are here as plain as they are amazing. The rocks are so hard that not only the fluting and polishing, but even the fine scratches which date back unnamable thousands of years are as evident as if they had been made yesterday. We may trace these evidences to a height of two thousand feet above the present valley bed. It is indubitable that an ice-river of this astounding depth once flowed through the vale of Hasli.

305. Yonder is the summit of the Siedelhorn; and if we gain it, the Unteraar glacier will lie like a map below us. From this commanding point we plainly see marked upon the mountain sides the height to which the ancient ice extended. The ice-ground part of the mountains is clearly distinguished from the splintered crests which in those distant days rose above the surface

of the glacier, and which must have then appeared as island peaks and crests in the midst of an ocean of ice.

366. We now scamper down the Siedelhorn, get once more into the valley of Hasli, along which we follow for more than twenty miles the traces of the ice. Fluted precipices, polished slabs, and beautifully-rounded granite domes. Right and left upon the mountain flanks, at great elevations, the evidences appear. We follow the footsteps of the glacier to the Lake of Brientz; and if we prolonged our enquiries, we should learn that all the lake beds of this region, at the time now referred to, bore the burden of immense masses of ice.

367. Instead of the vale of Hasli, we might take the valley of the Rhone. The traces of a mighty glacier, which formerly filled it, may be followed all the way to Martigny, which is 60 miles distant from the present ice. At Martigny the Rhone glacier was reinforced by another from Mont Blanc, and the welded masses moved onward, planing the mountains right and left, to the Lake of Geneva, the basin of which they entirely filled. Other evidences prove that the glacier did not end here, but pushed across the low country until it encountered the limestone barrier of the Jura Mountains.

§ 54. *Erratic Blocks.*

368. What are these other evidences? We have seen mighty rocks poised on the moraines of the Mer

de Glace, and we now know that, unless they are split
and shattered by the frost, these rocks will, at some
distant day, be landed bodily by the Glacier des Bois
in the valley of Chamouni. You have already learned
that these boulders often reveal the mineralogical nature
of the mountains among which the glacier has passed;
that specimens are thus brought down of a character
totally different from the rocks among which they are
finally landed; this is strikingly the case with the
erratic blocks stranded along the Jura.

369. For the Jura itself, as already stated, is limestone; there is no trace of native granite to be found.
amongst these hills. Still along the breast of the
mountain above the town of Neufchâtel, and at about
800 feet above the lake of Neufchâtel, we find stranded
a belt of granite boulders from Mont Blanc. And when
we clear the soil away from the adjacent mountain side,
we find upon the limestone rocks the scarrings of the
ancient glacier which brought the boulders here.

370. The most famous of these rocks, called the
Pierre à Bôt, measures 50 feet in length, 40 in height,
and 20 in width. Multiplying these three numbers
together, we obtain 40,000 cubic feet as the volume of
the boulder.

371. But this is small compared to some of the
rocks which constitute the freight of even recent
glaciers. Let us visit another of them. We have
already been to Stalden, where the valley divides into

two branches, the right branch running to St. Nicholas and Zermatt, and the left one to Saas and the Monte Moro. Three hours above Saas we come upon the end of the Allelein glacier, not filling the main valley, but thrown athwart it so as to stop its drainage like a dam. Above this ice-dam we have the Mattmark Lake, and at the head of the lake a small inn well known to travellers over the Monte Moro.

372. Close to this inn is the greatest boulder that we have ever seen. It measures 240,000 cubic feet. Looking across the valley we notice a glacier with its present end half a mile from the boulder. The stone, I believe, is serpentine, and were you and I to explore the Schwartzberg glacier to its upper fastnesses, we should find among them the birthplace of this gigantic stone. Four-and-forty years ago, when the glacier reached the place now occupied by the boulder, it landed there its mighty freight, and then retreated. There is a second ice-borne rock at hand which would be considered vast were it not dwarfed by the aspect of its huger neighbour.

373. Evidence of this kind might be multiplied to any extent. In fact, at this moment, distinguished men, like Professor Favre of Geneva, are determining from the distribution of the erratic blocks the extent of the ancient glaciers of Switzerland. It was, however, an engineer named Venetz that first brought these evidences to light, and announced to an incredulous world the

150 THE FORMS OF WATER IN

vast extension of the ancient ice. M. Agassiz afterwards developed and wonderfully expanded the discovery. Perhaps the most interesting observation
regarding ancient glaciers is that of Dr. Hooker,
who, during a recent visit to Palestine, found the
celebrated Cedars of Lebanon growing upon ancient
moraines.

§ 55. Ancient Glaciers of England, Ireland, Scotland, and Wales.

374. At the time the ice attained this extraordinary
development in the Alps, many other portions of Europe,
where no glaciers now exist, were covered with them.
In the Highlands of Scotland, among the mountains of
England, Ireland, and Wales, the ancient glaciers have
written their story as plainly as in the Alps themselves.
I should like to wander with you through Borrodale in
Cumberland, or through the valleys near Bethgellert in
Wales. Under all the beauty of the present scenery we
should discover the memorials of a time when the whole
region was locked in the embrace of ice. Professor
Ramsay is especially distinguished by his writings on
the ancient glaciers of Wales.

375. We have made the acquaintance of the Reeks
of Magillicuddy as the great condensers of Atlantic
vapour. At the time now referred to, this moisture
did not fall as soft and fructifying rain, but as snow,
which formed the nutriment of great glaciers. A chain

of lakes now constitutes the chief attraction of Killarney, the Lower, the Middle, and the Upper Lake. Let us suppose ourselves rowing towards the head of the Upper Lake with the Purple Mountain to our left. Remembering our travels in the Alps, you would infallibly call my attention to the planing of the rocks, and declare the action to be unmistakably that of glaciers. With our attention thus sharpened, we land at the head of the lake, and walk up the Black Valley to the base of Magillicuddy's Rocks. Your conclusion would be, that this valley tells a tale as wonderful as that of Hasli.

376. We reach our boat and row homewards along the Upper Lake. Its islands now possess a new interest for us. Some of them are bare, others are covered wholly or in part with luxuriant vegetation; but both the naked and clothed islands are glaciated. The weathering of ages has not altered their forms: there are the Cannon Rock, the Giant's Coffin, the Man of War, all sculptured as if the chisel had passed over them in our own lifetime. These lakes, now fringed with tender woodland beauty, were all occupied by the ancient ice. It has disappeared, and seeds from other regions have been wafted thither to sow the trees, the shrubs, the ferns, and the grasses which now beautify Killarney. Man himself, they say, has made his appearance in the world since that time of ice; but of the real period and manner of man's introduction little is professed to be known since, to make them square with science, new

meanings have been found for the beautiful myths and stories of the Bible.

377. It is the nature and tendency of the human mind to look backward and forward; to endeavour to restore the past and predict the future. Thus endowed, from data patiently and painfully won, we recover in idea a state of things which existed thousands, it may be millions, of years before the history of the human race began.

§ 56. The Glacial Epoch.

378. This period of ice-extension has been named the *Glacial Epoch*. In accounting for it great minds have fallen into grave errors, as we shall presently see.

379. The substance on which we have thus far been working exists in three different states: as a solid in ice; as a liquid in water; as a gas in vapour. To cause it to pass from one of these states to the next following one, *heat* is necessary.

380. Dig a hole in the ice of the Mer de Glace in summer, and place a thermometer in the hole; it will stand at 32° Fahr. Dip your thermometer into one of the glacier streams; it will still mark 32°. *The water is therefore as cold as ice.*

381. Hence the whole of the heat poured by the sun upon the glacier, and which has been absorbed by the glacier, is expended in simply liquefying the ice, and not in rendering either ice or water a single degree warmer.

382. Expose water to a fire; it becomes hotter for a time. It boils, and from that moment it ceases to get hotter. After it has begun to boil, all the heat communicated by the fire is carried away by the steam, *though the steam itself is not the least fraction of a degree hotter than the water.*

383. In fact, simply to liquefy ice a large quantity of heat is necessary, and to vaporize water a still larger quantity is necessary. And inasmuch as this heat does not render the water warmer than the ice, nor the steam warmer than the water, it was at one time supposed to be *hidden* in the water and in the steam. And it was therefore called *latent heat.*

384. Let us ask how much heat must the sun expend in order to convert a pound weight of the tropical ocean into vapour? This problem has been accurately solved by experiment. It would require in round numbers 1,000 times the amount of heat necessary to raise one pound of water one degree in temperature.

385. But the quantity of heat which would raise the temperature of a pound of water one degree would raise the temperature of a pound of iron *ten* degrees. This has been also proved by experiment. Hence to convert one pound of the tropical ocean into vapour the sun must expend 10,000 times as much heat as would raise one pound of iron one degree in temperature.

386. This quantity of heat would raise the temperature of 5 lbs. of iron 2,000 degrees, which is the fusing point of cast iron; at this temperature the metal would

not only be *white hot*, but would be passing into the
molten condition.

387. Consider the conclusions at which we have now
arrived. For every pound of tropical vapour, or for
every pound of Alpine ice produced by the congelation
of that vapour, an amount of heat has been expended
by the sun sufficient to raise 5 lbs. of cast iron to its
melting-point.

388. It would not be difficult to calculate approxi-
mately the weight of the Mer de Glace and its tributaries—to say, for example, that they contained so many
millions of millions of tons of ice and snow. Let the
place of the ice be taken by a mass of white-hot iron of
quintuple the weight; with such a picture before your
mind you get some notion of the enormous amount of
heat paid out by the sun to produce the present glacier.

389. You must think over this, until it is as clear as
sunshine. For you must never henceforth fall into the
error already referred to, and which has entangled so
many. So natural was the association of ice and cold,
that even celebrated men assumed that all that is needed
to produce a great extension of our glaciers is a diminution of the sun's temperature. Had they gone
through the foregoing reflections and calculations, they
would probably have demanded *more* heat instead of less
for the production of a 'glacial epoch.' What they
really needed were *condensers* sufficiently powerful to
congeal the vapour generated by the heat of the sun.

§ 57. *Glacier Theories.*

390. You have not forgotten, and hardly ever can forget, our climbs to the Cleft Station. Thoughts were then suggested which we have not yet discussed. We saw the branch glaciers coming down from their névés, welding themselves together, pushing through Trélaporte, and afterwards moving through the sinuous valley of the Mer de Glace. These appearances alone, without taking into account subsequent observations, were sufficient to suggest the idea that glacier ice, however hard and brittle it may appear, is really a viscous substance, resembling treacle, or honey, or tar, or lava.

§ 58. *Dilatation and Sliding Theories.*

391. Still this was not the notion expressed by the majority of writers upon glaciers. Scheuchzer of Zurich, a great naturalist, visited the glaciers in 1705, and propounded a theory of their motion. Water, he knew, expands in freezing, and the force of expansion is so great, that thick bombshells filled with water, and permitted to freeze, are, as we know (312), shattered to pieces by the ice within. Scheuchzer supposed that the water in the fissures of the glaciers, freezing there and expanding with resistless force, was the power which urged the glacier downwards. He added to this theory other notions of a less scientific kind.

392. Many years subsequently, De Charpentier of Bex renewed and developed this theory with such ability and completeness, that it was long known as Charpentier's Theory of Dilatation. M. Agassiz for a time espoused this theory, and it was also more or less distinctly held by other writers. The glacier, in fact, was considered to be a magazine of cold, capable of freezing all water percolating through it. The theory was abandoned when this notion of glacier cold was proved by M. Agassiz to be untenable.

393. In 1760, Altmann and Grüner propounded the view that glaciers moved by sliding over their beds. Nearly forty years subsequently, this notion was revived by De Saussure, and it has therefore been called 'De Saussure's Theory,' or the 'Sliding Theory' of glacier motion.

394. There was, however, but little reason to connect the name of De Saussure with this or any other theory of glaciers. Incessantly occupied in observations of another kind, this celebrated man devoted very little time or thought to the question of glacier motion. What he has written upon the subject reads less like the elaboration of a theory than the expression of an opinion.

§ 59. *Plastic Theory.*

395. By none of these writers is the property of viscosity or plasticity ascribed to glacier ice; the appearances of many glaciers are, however, so suggestive of

this idea that we may be sure it would have found more
frequent expression, were it not in such apparent con-
tradiction with our every-day experience of ice.

396. Still the idea found its advocates. In a little
book, published in 1773, and entitled ' Picturesque
Journey to the Glaciers of Savoy,' Bordier of Geneva
wrote thus :—' It is now time to look at all these objects
with the eyes of reason; to study, in the first place, the
position and the progression of glaciers, and to seek the
solution of their principal phenomena. At the first aspect of the ice-mountains an observation presents itself,
which appears sufficient to explain all. It is that the
entire mass of ice is connected together, and presses
from above downwards after the manner of fluids. Let
us then regard the ice, not as a mass entirely rigid
and immobile, but as a heap of coagulated matter, or as
softened wax, flexible and ductile to a certain point.' *
Here probably for the first time the quality of plasticity
is ascribed to the ice of glaciers.

397. To us, familiar with the aspect of the glaciers,
it must seem strange that this idea once expressed did
not at once receive recognition and development. But
in those early days explorers were few, and the ' Picturesque Journey' probably but little known, so that
the notion of plasticity lay dormant for more than half

* I am indebted to my distinguished friend Prof. Studer of Berne for
directing my attention to Bordier's book, and to my friends at the British
Museum for the great trouble they have taken to find it for me.

a century. But Bordier was at length succeeded by a
man of far greater scientific grasp and insight than
himself. This was Rendu, a Catholic priest and canon
when he wrote, and afterwards Bishop of Annecy. In
1841 Rendu laid before the Royal Academy of Sciences
of Savoy his 'Theory of the Glaciers of Savoy,' a con-
tribution for ever memorable in relation to this subject.*

398. Rendu seized the idea of glacier plasticity with
great power and clearness, and followed it resolutely to
its consequences. It is not known that he had ever
seen the work of Bordier; probably not, as he never
mentions it. Let me quote for you some of Rendu's
expressions, which, however, fail to give an adequate
idea of his insight and precision of thought :—' Between
the Mer de Glace and a river there is a resemblance so
complete that it is impossible to find in the glacier a
circumstance which does not exist in the river. In
currents of water the motion is not uniform either
throughout their width or throughout their depth.
The friction of the bottom and of the sides, with the
action of local hindrances, causes the motion to vary,
and only towards the middle of the surface do we
obtain the full motion.'

399. This reads like a prediction of what has since
been established by measurement. Looking at the
glacier of Mont Dolent, which resembles a sheaf in
form, wide at both ends and narrow in the middle, and

* 'Memoirs of the Academy,' vol. x.

reflecting that the upper wide part had become narrow,
and the narrow middle part again wide, Rendu observes,
'There is a multitude of facts which seem to necessitate
the belief that glacier ice enjoys a kind of ductility
which enables it to mould itself to its locality, to thin
out, to swell, and to contract as if it were a soft paste.'

400. To fully test his conclusions, Rendu required
the accurate measurement of glacier motion. Had he
added to his other endowments the practical skill of a
land-surveyor, he would now be regarded as the prince
of glacialists. As it was he was obliged to be content
with imperfect measurements. In one of his excursions he examined the guides regarding the successive
positions of a vast rock which he found upon the ice
close to the side of the glacier. The mean of five years
gave him a motion for this block of 40 feet a year.

401. Another block, the transport of which he subsequently measured more accurately, gave him a velocity
of 400 feet a year. Note his explanation of this dis-
crepancy:—'The enormous difference of these two observations arises from the fact that one block stood near
the centre of the glacier, which moves most rapidly,
while the other stood near the side, where the ice is
held back by friction.' So clear and definite were
Rendu's ideas of the plastic motion of glaciers, that had
the question of curvature occurred to him, I entertain
no doubt that he would have enunciated beforehand the

shifting of the point of maximum motion from side to
side across the axis of the glacier (§ 25).

402. It is right that you should know that scientific
men do not always agree in their estimates of the com-
parative value of facts and ideas; and it is especially
right that you should know that your present tutor
attaches a very high value to ideas when they spring
from the profound and persistent pondering of superior
minds, and are not, as is too often the case, thrown out
without the warrant of either deep thought or natural
capacity. It is because I believe Rendu's labours fulfil
this condition, that I ascribe to them so high a value.
But when you become older and better informed, you
may differ from me; and I write these words lest you
should too readily accept my opinion of Rendu. Judge
me, if you care. to do so, when your knowledge is
matured. I certainly shall not fear your verdict.

403. But, much as I prize the prompting idea, and
thoroughly as I believe that often in it the force of
genius mainly lies, it would, in my opinion, be an error
of omission of the gravest kind, and which, if habitual,
would ensure the ultimate decay of natural knowledge,
to neglect verifying our ideas, and giving them outward
reality and substance when the means of doing so are
at hand. In science thought, as far as possible, ought
to be wedded to fact. This was attempted by Rendu,
and in great part accomplished by Agassiz and Forbes.

§ 60. *Viscous Theory.*

404. Here indeed the merits of the distinguished glacialist last named rise conspicuously to view. From the able and earnest advocacy of Professor Forbes, the public knowledge of this doctrine of glacial plasticity is almost wholly derived. He gave the doctrine a more distinctive form; he first applied the term *viscous* to glacier ice, and sought to found upon precise measurements a 'Viscous Theory' of glacier motion.

405. I am here obliged to state facts in their historic sequence. Professor Forbes when he began his investigations was acquainted with the labours of Rendu. In his earliest work upon the Alps he refers to those labours in terms of flattering recognition. But though as a matter of fact Rendu's ideas were there to prompt him, it would be too much to say that he needed their inspiration. Had Rendu not preceded him, he might none the less have grasped the idea of viscosity, executing his measurements and applying his knowledge to maintain it. Be that as it may, the appearance of Professor Forbes on the Unteraar glacier in 1841, and on the Mer de Glace in 1842, and his labours then and subsequently, have given him a name not to be forgotten in the scientific history of glaciers.

406. The theory advocated by Professor Forbes was enunciated by himself in these words :—' A glacier is an imperfect fluid, or viscous body, which is urged down

M

slopes of certain inclination by the natural pressure of its parts.' In 1773 Bordier wrote thus :—' As the glaciers always advance upon the plain, and never disappear, it is absolutely essential that new ice shall perpetually take the place of that which is melted : it must therefore be pressed forward from above. One can hardly refuse then to accept the astonishing truth, that this vast extent of hard and solid ice moves as a single piece downwards.' In the passage already quoted he speaks of the ice being pressed as a fluid from above. These constitute, I believe, Bordier's contributions to this subject. The quotations show his sagacity at an early date; but, in point of completeness, his views are not to be compared with those of Rendu and Forbes.

407. I must not omit to state here that though the idea of viscosity has not been espoused by M. Agassiz, his measurements, and maps of measurements, on the Unteraar glacier have been recently cited as the most clear and conclusive illustrations of a quality which, at all events, closely resembles viscosity.

408. But why, with proofs before him more copious and characteristic than those of any other observer, does M. Agassiz hesitate to accept the idea of viscosity as applied to ice? Doubtless because he believes the notion to be contradicted by our every-day experience of the substance.

409. Take a mass of ice ten or even fifteen cubic feet

in volume; draw a saw across it to a depth of half an inch or an inch; and strike a pointed pricker, not thicker than a very small round file, into the groove; the substance will split from top to bottom with a clean crystalline fracture. How is this brittleness to be reconciled with the notion of viscosity?

410. We have, moreover, been upon the glacier and have witnessed the birth of crevasses. We have seen them beginning as narrow cracks suddenly formed, days being required to open them a single inch. In many glaciers fissures may be traced narrow and profound for hundreds of yards through the ice. What does this prove? Did the ice possess even a very small modicum of that power of stretching, which is characteristic of a viscous substance, such crevasses could not be formed.

411. Still it is undoubted that the glacier moves like a viscous body. The centre flows past the sides, the top flows over the bottom, and the motion through a curved valley corresponds to fluid motion. Mr. Mathews, Mr. Froude, and above all Signor Bianconi, have, more-over, recently made experiments on ice which strikingly illustrate the flexibility of the substance. These experiments merit, and will doubtless receive, full attention at a future time.

§ 61. Regulation Theory.

412. I will now describe to you an attempt that has been made of late years to reconcile the brittleness of

ice with its motion in glaciers. It is founded on the
observation, made by Mr. Faraday in 1850, that when
two pieces of thawing ice are placed together they
freeze together at the place of contact.

413. This fact may not surprise you; still it surprised
Mr. Faraday and others, and men of very great distinc-
tion in science have differed in their interpretation of the
fact. The difficulty is to explain where, or how, in ice
already thawing the cold is to be found requisite to freeze
the film of water between the two touching surfaces.

414. The word *Regelation* was proposed by Dr. Hooker
to express the freezing together of two pieces of thawing
ice observed by Faraday; and the memoir in which the
term was first used was published by Mr. Huxley and
Mr. Tyndall in the Philosophical Transactions for 1857.

415. The *fact* of regelation, and its application irre-
spective of the *cause* of regelation, may be thus illustrated :—Saw two slabs from a block of ice, and bring
their flat surfaces into contact; they immediately freeze
together. Two plates of ice, laid one upon the other,
with flannel round them overnight, are sometimes so
firmly frozen in the morning that they will rather break
elsewhere than along their surface of junction. If you
enter one of the dripping ice-caves of Switzerland, you
have only to press for a moment a slab of ice against
the roof of the cave to cause it to freeze there and
stick to the roof.

416. Place a number of fragments of ice in a basin of

water, and cause them to touch each other; they freeze
together where they touch.   You can form a chain of
such fragments; and then, by taking hold of one end
of the chain, you can draw the whole series after it.
Chains of icebergs are sometimes formed in this way
in the Arctic seas.

417. Consider what follows from these observations.
Snow consists of small particles of ice.   Now if by
pressure we squeeze out the air entangled in thawing
snow, and bring the little ice-granules into close contact,
they may be expected to freeze together; and if the
expulsion of the air be complete, the squeezed snow may
be expected to assume the appearance of compact ice.

418. We arrive at this conclusion by reasoning; let
us now test it by experiment, employing a suitable hydraulic press, and a mould to hold the snow.   In exact
accordance with our expectation, we convert by pressure
the snow into ice.*

419. Place a compact mass of ice in a proper mould,
and subject it to pressure.   It breaks in pieces: squeeze
the pieces forcibly together; they re-unite by regelation, and a compact piece of ice, totally different in
shape from the first one, is taken from the press.   To
produce this effect the ice must be in a thawing condition.   When its temperature is much below the
melting point it is crushed by pressure, not into a

* A similar experiment was made by the Messrs. Schlagintweit prior to
the discovery which explains it, and which therefore remained unsolved.

pellucid mass of another shape, but into a white
powder.

420. By means of suitable moulds you may in this
way change the shape of ice to any extent, turning out
spheres, and cups, and rings, and twisted ropes of the
substance; the change of form in these cases being
effected through rude fracture and regelation.

421. By applying the pressure carefully, rude fracture
may be avoided, and the ice compelled slowly to change
its form as if it were a plastic body.

422. Now our first experiment illustrates the consolidation of the snows of the higher Alpine regions.
The deeper layers of the névé have to bear the weight
of all above them, and are thereby converted into more
or less perfect ice. And our last experiment illustrates
the changes of form observed upon the glacier, where,
by the slow and constant application of pressure, the ice
gradually moulds itself to the valley, which it fills.

423. In glaciers, however, we have also ample illustrations of rude fracture and regelation. The opening
and closing of crevasses illustrate this. The glacier
is broken on the cascades and mended at their bases.
When two branch glaciers lay their sides together, the
regelation is so firm that they begin immediately to
flow in the trunk glacier as a single stream. The
medial moraine gives no indication by its slowness of
motion that it is derived from the sluggish ice of the
sides of the branch glaciers.

424. The gist of the Regelation Theory is that the ice of glaciers changes its form and preserves its continuity under *pressure* which keeps its particles together. But when subjected to *tension*, sooner than stretch it *breaks*, and behaves no longer as a viscous body.

§ 62. *Cause of Regelation.*

425. Here the fact of regelation is applied to explain the plasticity of glacier ice, no attempt being made to assign the cause of regelation itself. They are two entirely distinct questions. But a little time will be well spent in looking more closely into the cause of regelation. You may feel some surprise that eminent men should devote their attention to so small a point, but we must not forget that in nature nothing is small. Laws and principles interest the scientific student most, and these may be as well illustrated by small things as by large ones.

426. The question of regelation immediately connects itself with that of ' latent heat,' already referred to, (383) but which we must now subject to further examination. To melt ice, as already stated, a large amount of heat is necessary, and in the case of the glaciers this heat is furnished by the sun. Neither the ice so melted nor the water which results from its liquefaction can fall below 32° Fahrenheit. The freezing point of water and the melting point of ice touch each other, as it were, at this temperature. A hair's-breadth lower water freezes; a hair's-breadth higher ice melts.

427. But if the ice could be caused to melt without this supply of solar heat, a temperature lower than that of ordinary thawing ice would result. When snow and salt, or pounded ice and salt, are mixed together, the salt causes the ice to melt, and in this way a cold of 20 or 30 degrees below the freezing point may be produced. Here, in fact, the ice consumes *its own warmth* in the work of liquefaction. Such a mixture of ice and salt is called ' a freezing mixture.'

428. And if by any other means ice at the temperature of 32° Fahrenheit could be liquefied without access of heat from without, the water produced would be colder than the ice. Now Professor James Thomson has proved that ice may be liquefied by mere *pressure*, and his brother, Sir William Thomson, has also shown that water under pressure requires a lower temperature to freeze it than when the pressure is removed. Professor Mousson subsequently liquefied large masses of ice by a hydraulic press; and by a beautiful experiment Professor Helmholtz has proved that water in a vessel from which the air has been removed, and which is therefore relieved from the pressure of the atmosphere, freezes and forms ice-crystals when surrounded by melting ice. All these facts are summed up in the brief statement *that the freezing point of water is lowered by pressure.* *

429. For our own instruction we may produce the

* Professor James Thomson and Professor Clausius proved this independently and almost contemporaneously.

liquefaction of ice by pressure in the following way:
—You remember the beautiful flowers obtained when a
sunbeam is sent through lake ice (§ 11), and you have
not forgotten that the flowers always form parallel to
the surface of freezing. Let us cut a prism, or small
column of ice with the planes of freezing running across
it at right angles; we place that prism between two
slabs of wood, and bring carefully to bear upon it the
squeezing force of a small hydraulic press.

430. It is well to converge by means of a concave mirror
a good light upon the ice, and to view it through a
magnifying lens. You already see the result. Hazy
surfaces are formed in the very body of the ice, which
gradually expand as the pressure is slowly augmented.
Here and there you notice something resembling crystallisation; fern-shaped figures run with considerable
rapidity through the ice, and when you look carefully at
their points and edges you find them in visible motion.
These hazy surfaces are spaces of liquefaction, and the
motion you see is that of the ice falling to water under
the pressure. That water is colder than the ice was
before the pressure was applied, and if the pressure be
relieved, not only does the liquefaction cease, but the
water re-freezes. The cold produced by its liquefaction
under pressure is sufficient to re-congeal it when the
pressure is removed.

431. If instead of diffusing the pressure over surfaces of considerable extent, we concentrate it on a

small surface, the liquefaction will of course be more rapid, and this is what Mr. Bottomley has recently done in an experiment of singular beauty and interest. Let us support on blocks of wood the two ends of a bar of ice 10 inches long, 4 inches deep, and 3 wide, and let us loop over its middle a copper wire one-twentieth, or even one-tenth, of an inch in thickness. Connecting the two ends of the wire together, and suspending from it a weight of 12 or 14 pounds, the whole pressure of this weight is concentrated on the ice which supports the wire. What is the consequence? The ice underneath the wire liquefies; the water of liquefaction escapes round the wire, but the moment it is relieved from the pressure it freezes, and round about the wire, even before it has entered the ice, you have a frozen casing. The wire continues to sink in the ice; the water incessantly escapes, freezing as it does so behind the wire. In half an hour the weight falls; the wire has gone clean through the ice. You can plainly see where it has passed, but the two severed pieces of ice are so firmly frozen together that they will break elsewhere as soon as along the surface of regelation.

432. Another beautiful experiment bearing upon this point has recently been made by M. Boussingault. He filled a hollow steel cylinder with water and chilled it. In passing to ice water, as you know, expands (§ 45); in fact, room for expansion is a necessary condition of solidification. But in the present case the strong steel

resisted the expansion, the water in consequence remaining liquid at a temperature of more than 30° Fahr. below the ordinary freezing point. A bullet within the cylinder rattled about at this temperature, showing that the water was still liquid. On opening the tap the liquid, relieved of the pressure, was instantly converted into ice.

433. It is only substances which *expand* on solidifying that behave in this manner. The metal bismuth, as we know, is an example similar to water; while lead, wax, or sulphur, all of which contract on solidifying, have their point of fusion *heightened* by pressure.

434. And now you are prepared to understand Professor James Thomson's theory of regelation. When two pieces of ice are pressed together liquefaction, he contends, results. The water spreads out around the points of pressure, and when released re-freezes, thus forming a kind of cement between the pieces of ice.

§ 63. *Faraday's View of Regelation.*

435. Faraday's view of regelation is not so easily expressed, still I will try to give you some notion of it, dealing in the first place with admitted facts. Water, even in open vessels, may be lowered many degrees below its freezing temperature, and still remain liquid; it may also be raised to a temperature far higher than its boiling point, and still resist boiling. This is due to the mutual cohesion of the water particles, which resists the change

of the liquid either into the solid or the vaporous condition.

436. But if into the over-chilled water you throw a particle of ice, the cohesion is ruptured, and congelation immediately sets in. And if into the superheated water you introduce a bubble of air or of steam, cohesion is likewise ruptured, and ebullition immediately commences.

437. Faraday concluded that *in the interior* of any body, whether solid or liquid, where every particle is grasped so to speak by the surrounding particles, and grasps them in turn, the bond of cohesion is so strong as to require a higher temperature to change the state of aggregation than is necessary *at the surface*. At the surface of a piece of ice, for example, the molecules are free on one side from the control of other molecules; and they therefore yield to heat more readily than in the interior. The bubble of air or steam in overheated water also frees the molecules on one side; hence the ebullition consequent upon its introduction. Practically speaking, then, the point of liquefaction of the interior ice is higher than that of the superficial ice. Faraday also refers to the special solidifying power which bodies exert upon their own molecules. Camphor in a glass bottle fills the bottle with an atmosphere of camphor. In such an atmosphere large crystals of the substance may grow by the incessant deposition of camphor molecules upon camphor, at a temperature

too high to permit of the slightest deposit *upon the adjacent glass.* A similar remark applies to sulphur, phosphorus, and the metals in a state of fusion. They are deposited upon solid portions of their own substance at temperatures not low enough to cause them to solidify against other substances.

438. Water furnishes an eminent example of this special solidifying power. It may be cooled ten degrees and more below its freezing point without freezing. But this is not possible if the smallest fragment of ice be floating in the water. It then freezes accurately at 32° Fahr., depositing itself, however, not upon the sides of the containing vessel, but *upon the ice.* Faraday observed in a freezing apparatus thin crystals of ice growing in ice-cold water to a length of six, eight, or ten inches, at a temperature incompetent to produce their deposition upon the sides of the containing vessel.

439. And now we are prepared for Faraday's view of regelation. When the surfaces of two pieces of ice, covered with a film of the water of liquefaction, are brought together, the covering film is transferred from the surface to the centre of the ice, where the point of liquefaction, as before shown, is higher than at the surface. The special solidifying power of ice upon water is now brought into play *on both sides of the film.* Under these circumstances, Faraday held that the film would congeal, and freeze the two surfaces together.

440. The lowering of the freezing point by pressure amounts to no more than one-seventieth of a degree Fahrenheit for a whole atmosphere. Considering the infinitesimal fraction of this pressure which is brought into play in some cases of regelation, Faraday thought its effect insensible. He suspended pieces of ice, and brought them into contact without sensible pressure, still they froze together. Professor James Thomson, however, considered that even the capillary attraction exerted between two such masses would be sufficient to produce regelation. You may make the following experiments, in further illustration of this subject :—

441. Place a small piece of ice on water, and press it underneath the surface by a second piece. The submerged piece may be so small as to render the pressure infinitesimal; still it will freeze to the under surface of the superior piece.

442. Place two pieces of ice in a basin of warm water, and allow them to come together; they freeze together when they touch. The parts surrounding the place of contact melt away, but the pieces continue for a time united by a narrow bridge of ice. The bridge finally melts, and the pieces for a moment are separated. But capillary attraction immediately draws them together, and regelation sets in once more. A new bridge is formed, which in its turn is dissolved, the separated pieces again closing up. A kind of pulsation is thus established between the two pieces of ice. They touch,

they freeze, a bridge is formed and melted; and thus the rhythmic action continues until the ice disappears.

443. According to Professor James Thomson's theory, pressure is necessary to liquefy the ice. The heat necessary for liquefaction must be drawn from the ice itself, and the cold water must escape from the pressure to be re-frozen. Now in the foregoing experiments the cold water, instead of being allowed to freeze, *issues into the warm water*, still the floating fragments regelate in a moment. The touching surfaces may, moreover, be convex; they may be reduced practically to *points*, clasped all round by the warm water, which indeed rapidly dissolves them as they approach each other; still they freeze immediately when they touch.

444. You may learn from this discussion that in scientific matters, as in all others, there is room for differences of opinion. The frame of mind to be cultivated here is a suspension of judgment as long as the meaning remains in doubt. It may be that Faraday's action and Thomson's action come both into play. I cannot do better than finish these remarks by quoting Faraday's own concluding words, which show how in his mind scientific conviction dwelt apart from dogmatism :—'No doubt,' he says, 'nice experiments will enable us hereafter to criticise such results as these, and separating the true from the untrue will establish the correct theory of regelation.'

§ 64. *The Blue Veins of Glaciers.*

445. We now approach the end, one important question only remaining to be discussed. Hitherto we have kept it back, for a wide acquaintance with the glaciers was necessary to its solution. We had also to make ourselves familiar by actual experiment with the power of ice, softened by thaw, to yield to pressure, and to liquefy under such pressure.

446. Snow is white. But if you examine its individual particles you would call them *transparent*, not white. The whiteness arises from the mixture of the ice particles with small spaces of air. In the case of all transparent bodies whiteness results from such a mixture. The clearest glass or crystal when crushed becomes a white powder. The foam of champagne is white through the intimate admixture of a transparent liquid with transparent carbonic acid gas. The whitest paper, moreover, is composed of fibres which are individually transparent.

447. It is not, however, the air or the gas, but the *optical severance* of the particles, giving rise to a multitude of reflexions of the white solar light at their surfaces, that produces the whiteness.

448. The whiteness of the surface of a clean glacier (112), and of the icebergs of the Märgelin See (357), has been already referred to a similar cause. The surface is broken into innumerable fissures by the solar

heat, the reflexion of solar light from the sides of the
little fissures producing the observed appearance.

449. In like manner if you freeze water in a test-
tube by plunging it into a freezing mixture, the ice
produced is white. For the most part also the ice
formed in freezing machines is white. Examine such
ice, and you will find it filled with small air-bubbles.
When the freezing is extremely slow the crystallising
force pushes the air effectually aside, and the result-
ing ice is transparent; when the freezing is rapid, the
air is entangled before it can escape, and the ice is
translucent. But even in the case of quick freezing
Mr. Faraday obtained transparent ice by skilfully re-
moving the air-bubbles as fast as they appeared with
a feather.

450. In the case of lake ice the freezing is not uniform, but intermittent. It is sometimes slow, sometimes
rapid. When slow the air dissolved in the water is
effectually squeezed out and forms a layer of bubbles on
the under-surface of the ice. An act of sudden freezing
entangles this air, and hence we find lake ice usually
composed of layers alternately clear, and filled with
bubbles. Such layers render it easy to detect the planes
of freezing in lake ice.

451. And now for the bearing of these facts. Under
the fall of the Géant, at the base of the Talèfre cascade,
and lower down the Mer de Glace; in the higher regions of the Grindelwald, the Aar, the Aletsch and the

N

Görner glaciers, the ice does not possess the transparency which it exhibits near the ends of the glaciers. It is white, or whitish. Why? Examination shows it to be filled with small air-bubbles; and these, as we now learn, are the cause of its whiteness.

452. They are the residue of the air originally entangled in the snow, and connected, as before stated, with the whiteness of the snow. During the descent of the glacier, the bubbles are gradually expelled by the enormous pressures brought to bear upon the ice. Not only is the expulsion caused by the mechanical yielding of the soft thawing ice, but the liquefaction of the substance at places of violent pressure, opening, as it does, fissures for the escape of the air, must play an important part in the consolidation of the glacier.

453. The expulsion of the bubbles is, however, not uniform; for neither ice nor any other substance offers an absolutely uniform resistance to pressure. At the base of every cascade that we have visited, and on the walls of the crevasses there formed, we have noticed innumerable blue streaks drawn through the white translucent ice, and giving the whole mass the appearance of lamination. These blue veins turned out upon examination to be spaces from which the air-bubbles had been almost wholly expelled, translucency being thus converted into transparency.

454. This is the *veined* or *ribboned structure* of glaciers, regarding the origin of which diverse opinions are now entertained.

455. It is now our duty to take up the problem, and to solve it if we can. On the névés of the Col du Géant, and other glaciers, we have found great cracks, and faults, and *Bergschrunds*, exposing deep sections of the névé; and on these sections we have found marked the edges of half-consolidated strata evidently produced by successive falls of snow. The névé is stratified because its supply of material from the atmosphere is intermittent, and when we first observed the blue veins we were disposed to regard them as due to this stratification.

456. But observation and reflexion soon dispelled this notion. Indeed it could hardly stand in the presence of the single fact that at the bases of the ice-falls the veins are always *vertical*, or nearly so. We saw no way of explaining how the horizontal strata of the néve could be so tilted up at the base of the fall as to be set on edge. Nor is the aspect of the veins that of stratification.

457. On the central portions of the cascades, moreover, there are no signs of the veins. At the bases they first appear, reaching in each case their maximum development a little below the base. As you and I stood upon the heights above the Zäsenberg and scrutinised the cascade of the 'Strahleck branch of the Grindelwald glacier, we could not doubt that the base of the fall was the birthplace of the veins. We called this portion of the glacier a 'Structure Mill,' intimating

that here, and not on the névé, the veined structure was manufactured.

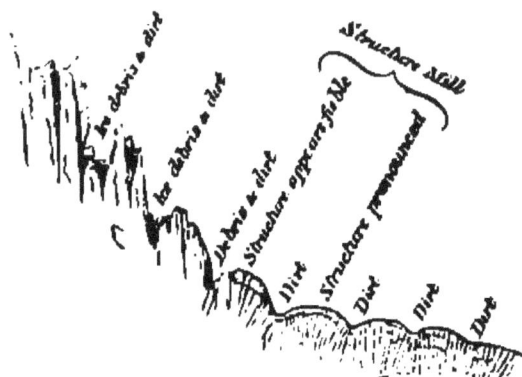

SECTION OF ICEFALL, AND GLACIER BELOW IT, SHOWING ORIGIN OF VEINED STRUCTURE.

458. This, however, is, at bottom, the language of strong *opinion* merely, not that of *demonstration*; and in science opinion ought to content us only so long as positive proof is unattainable. The love of repose must not prevent us from seeking this proof. There is no sterner conscience than the scientific conscience, and it demands, in every possible case, the substitution for private conviction of demonstration which shall be conclusive to all.

459. Let us, for example, be shown a case in which the stratification of the névé is prolonged into the glacier; let us see the planes of bedding and the planes of lamination existing side by side, and still indubitably

distinct. Such an observation would effectually exclude stratification from the problem of the veined structure, and through the removal of this tempting source of error, we should be rendered more free to pursue the truth.

460. We sought for this conclusive test upon the Mer de Glace, but did not find it. We sought it on the Grindelwald, and the Aar glaciers,* with an equal want of success. On the Aletsch glacier, for the first time, we observed the apparent coexistence of bedding and structure, the one *cutting* the other upon the walls of the same crevasse. Still the case was not sufficiently pronounced to produce entire conviction, and we visited the Görner glacier with the view of following up our quest.

STRUCTURE AND BEDDING ON ALETSCH GLACIER.

461. Here day after day added to the conviction that the bedding and the structure were two different things.

* M. Agassiz, however, reports a case of the kind upon the glacier of the Aar.

Still day after day passed without revealing to us the final proof. Surely we have not let our own ease stand in the way of its attainment, and if we retire baffled we shall do so with the consciousness of having done our best. Yonder, however, at the base of the Matterhorn, is the Furgge glacier that we have not yet explored. Upon it our final attempt must be made.

462. We get upon the glacier near its end, and ascend it. We are soon fronted by a barrier composed of three successive walls of névé, the one rising above the other, and each retreating behind the other. The bottom of each wall is separated from the top of the succeeding one by a ledge, on which threatening masses of broken névé now rest. We stand amid blocks and rubbish which have been evidently discharged from these ledges, on which other masses, ready apparently to tumble, are now poised.

463. On the vertical walls of this barrier we see, marked with the utmost plainness, the horizontal lines of stratification, while something exceedingly like the veined structure appears to cross the lines of bedding at nearly a right angle. The vertical surface is, however weathered, and the lines of structure, if they be such, are indistinct. The problem now is to remove the surface, and expose the ice underneath. It is one of the many cases that have come before us, where the value of an observation is to be balanced against the danger which it involves.

464. We do nothing rashly; but, scanning the ledges and selecting a point of attack, we conclude that the danger is not too great to be incurred. We advance to the wall, remove the surface, and are rewarded by the discovery underneath it of the true blue veins. They, moreover, are vertical, while the bedding is horizontal. Bruce, as you know, was defeated in many a battle, but he persisted and won at last. Here, upon the Furgge glacier, you also have fought and won your little Bannockburn.

STRUCTURE AND BEDDING OF FURGGE GLACIER.

465. But let us not use the language of victory too soon. The stratification theory has been removed out of the field of explanation, but nothing has as yet been offered in its place.

§ 65. *Relation of Structure to Pressure.*

466. This veined structure was first described by the distinguished Swiss naturalist, Guyot, now a resident in

the United States. From the Grimsel Pass I have
already pointed out to you the Gries glacier overspreading the mountains at the opposite side of the
valley of the Rhone. It was on this glacier that
M. Guyot made his observation.

467. 'I saw,' he said, 'under my feet the surface of
the entire glacier covered with regular furrows, from
one to two inches wide, hollowed out in a half-snowy
mass, and separated by protruding plates of harder and
more transparent ice. It was evident that the glacier
here was composed of two kinds of ice, one that of the
furrows, snowy and more easily melted; the other of
the plates, more perfect, crystalline, glassy, and resistant; and that the unequal resistance which the two
kinds of ice presented to the atmosphere was the cause
of the ridges.

468. 'After having followed them for several hundred yards, I reached a crevasse twenty or thirty feet
wide, which, as it cut the plates and furrows at right
angles, exposed the interior of the glacier to a depth of
thirty or forty feet, and gave a beautiful transverse
section of the structure. As far as my eyes could reach,
I saw the mass of the glacier composed of layers of
snowy ice, each two of which were separated by one of
the hard plates of which I have spoken, the whole
forming a regularly laminated mass, which resembled
certain calcareous slates.'

469. I have not failed to point out to you upon all

the glaciers that we have visited the little superficial
furrows here described; and you have, moreover,
noticed that in the furrows mainly is lodged the finer
dirt which is scattered over the glacier. They suggest the passage of a rake over the ice. And when-
ever these furrows were interrupted by a crevasse, the
veined structure invariably revealed itself upon the
walls of the fissure. The surface grooving is indeed
an infallible indication of the interior lamination of
the ice.

470. We have tracked the structure through the
various parts of the glaciers at which its appearance
was most distinct; and we have paid particular attention to the condition of the ice at these places. The
very fact of its cutting the crevasses at right angles is
significant. We know the mechanical origin of the
crevasses; that they are cracks formed at right angles
to lines of tension. But since the crevasses are also
perpendicular to the planes of structure, these planes
must be parallel to the lines of tension.

471. On the glaciers, however, tension rarely occurs
alone. At the sides of the glacier, for example, where
marginal crevasses are formed, the tension is always
accompanied by pressure; the one force acting at right
angles to the other. Here, therefore, the veined
structure, which is parallel to the lines of tension, *is
perpendicular to the lines of pressure.*

472. That this is so will be evident to you in a

moment. Let the adjacent figure represent the channel
of the glacier moving in the direction of the arrow.
Suppose three circles to be marked upon the ice, one at
the centre and the two others at the sides. In a glacier

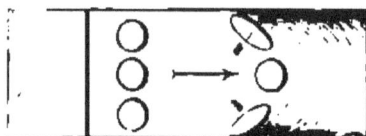

of uniform inclination all these circles would move
downward, the central one only remaining a circle. By
the retardation of the sides the marginal circles would
be drawn out to ovals. The two circles would be *elon-
gated* in one direction, and *compressed* in another.
Across the long diameter, which is the direction of
strain, we have the marginal crevasses ; across the short
diameter *m n*, which is the direction of pressure, we have
the *marginal veined structure.*

473. This association of pressure and structure is
invariable. At the bases of the cascades, where the
inclination of the bed of the glacier suddenly changes,
the pressure in many cases suffices not only to close the
crevasses but to violently squeeze the ice. At such
places the structure always appears, sweeping quite
across the glacier. When two branch glaciers unite,
their mutual thrust intensifies the pre-existing marginal
structure of the branches, and developes new planes of
lamination. Under the medial moraines, therefore, we

have usually a good development of the structure. It
is finely displayed, for example, under the great medial
moraine of the glacier of the Aar.

474. Upon this glacier, indeed, the blue veins were
observed independently three years after M. Guyot had
first described them. I say independently, because M.
Guyot's description, though written in 1838, remained
unprinted, and was unknown in 1841 to the observers on
the Aar. These were M. Agassiz and Professor Forbes.
To the question of structure Professor Forbes subsequently devoted much attention, and it was mainly
his observations and reasonings that gave it the important position now assigned to it in the phenomena
of glaciers.

475. Thus without quitting the glaciers themselves, we
establish the connexion between pressure and structure.
Is there anything in our previous scientific experience
with which these facts may be connected? The new
knowledge of nature must always strike its roots into
the old, and spring from it as an organic growth.

§ 66. *Slate Cleavage and Glacier Lamination.*

476. M. Guyot threw out an exceedingly sagacious
hint, when he compared the veined structure to the
cleavage of slate rocks. We must learn something of
this cleavage, for it really furnishes the key to the
problem which now occupies us. Let us go then to the

quarries of Bangor or Cumberland, and observe the ·
quarrymen in their sheds splitting the rocks. With a
sharp point struck skilfully into the edge of the slate,
they cause it to divide into thin plates, fit for roofing
or ciphering, as the case may be. The surfaces along
which the rock cleaves are called its *planes of cleavage*.

477. All through the quarry you notice the direction
of these planes to be perfectly constant. How is this
laminated structure to be accounted for?

478. You might be disposed to consider that cleavage
is a case of stratification or bedding; for it is true that
in various parts of England there are rocks which can
be cloven into thin flags along the planes of bedding.
But when we examine these slate rocks we verify the
observation, first I believe made by the eminent and
venerable Professor Sedgwick, that the planes of bedding usually run across the planes of cleavage.

479. We have here, as you observe, a case exactly
similar to that of glacier lamination, which we were
at first disposed to regard as due to stratification. We
afterwards, however, found planes of lamination crossing
the layers of the névé, exactly as the planes of cleavage
cross the beds of slate rocks.

480. But the analogy extends further. Slate cleavage
continued to be a puzzle to geologists till the late Mr.
Daniel Sharpe made the discovery that shells and other
fossils and bodies found in slate rocks are invariably
flattened out in the planes of cleavage.

481. Turn into any well-arranged museum—for example, into the School of Mines in Jermyn Street, and observe the evidence there collected. Look particularly to the fossil trilobites taken from the slate rock. They are in some cases squeezed to one third of their primitive thickness. Numerous other specimens show in the most striking manner the flattening out of shells.

482. To the evidence adduced by Mr. Sharpe, Mr. Sorby added other powerful evidence, founded upon the microscopic examination of slate rock. Taking both into account, the conclusion is irresistible that such rocks have suffered enormous pressure at right angles to the planes of cleavage, exactly as the glacier has demonstrably suffered great pressure at right angles to its planes of lamination.

483. The association of pressure and cleavage is thus demonstrated; but the question arises, do they stand to each other in the relation of cause and effect? The only way of replying to this question is to combine artificially the conditions of nature, and see whether we cannot produce her results.

484. The substance of slate rocks was once a plastic mud, in which fossils were embedded. Let us imitate the action of pressure upon such mud by employing, instead of it, softened white wax. Placing a ball of the wax between two glass plates, wetted to prevent it from sticking, we apply pressure and flatten out the wax.

485. The flattened mass is at first too soft to cleave

sharply; but you can see, by tearing, that it is laminated. Let us chill it with ice. We find afterwards
that no slate rock ever exhibited so fine a cleavage.
The laminæ, it need hardly be said, are perpendicular
to the pressure.

486. One cause of this lamination is that the wax is
an aggregate of granules the surfaces of which are places
of weak cohesion; and that by the pressure these
granules are squeezed flat, thus producing planes of
weakness at right angles to the pressure.

487. But the main cause of the cleavage 1 take to be
the lateral sliding of the particles of wax over each
other. Old attachments are thereby severed, which
the new ones fail to make good. Thus the tangential
sliding produces lamination, as the rails near a station
are caused to exfoliate by the gliding of the wheel.

488. Instead of wax we may take the slate itself,
grind it to fine powder, add water, and thus reproduce
the pristine mud. By the proper compression of such
mud, in one direction, the cleavage is restored.

489. Call now to mind the evidences we have had
of the power of thawing ice to yield to pressure.
Recollect the shortening of the Glacier du Géant, and
the squeezing of the Glacier de Léchaud, at Trélaporte.
Such a substance, slowly acted upon by pressure, will
yield laterally. Its particles will slide over each other,
the severed attachments being immediately made good
by regelation. It will not yield uniformly, but along

special planes. It will also liquefy, not uniformly, but along special surfaces. Both the sliding and the liquefaction will take place principally at right angles to the pressure, and glacier lamination is the result.

490. As long as it is sound the laminated glacier ice resists cleavage. Regelation, as I have said, makes the severed attachments good. But when such ice is exposed to the weather the structure is revealed, and the ice can then be cloven into tablets a square foot, or even a square yard in area.

§ 67. *Conclusion.*

491. Here, my friend, our labours close. It has been a true pleasure to me to have you at my side so long. In the sweat of our brows we have often reached the heights where our work lay, but you have been steadfast and industrious throughout, using in all possible cases your own muscles instead of relying upon mine. Here and there I have stretched an arm and helped you to a ledge, but the work of climbing has been almost exclusively your own. It is thus that I should like to teach you all things; showing you the way to profitable exertion, but leaving the exertion to you—more anxious to bring out your manliness in the presence of difficulty than to make your way smooth by toning difficulties down.

492. Steadfast, prudent, without terror, though not

at all times without awe, I have found you on rock and ice, and you have shown the still rarer quality of steadfastness in intellectual effort. As here set forth, our task seems plain enough, but you and I know how often we have had to wrangle resolutely with the facts to bring out their meaning. The work, however, is now done, and you are master of a fragment of that sure and certain knowledge which is founded on the faithful study of nature. Is it not worth the price paid for it? Or rather, was not the paying of the price—the healthful, if sometimes hard, exercise of . mind and body, upon alp and glacier—a portion of our delight?

493. Here then we part. And should we not meet again, the memory of these days will still unite us. Give me your hand. Good bye.

www.ingramcontent.com/pod-product-compliance
Lightning Source LLC
Chambersburg PA
CBHW021709210326
41599CB00013B/1586